PRINTING APPARATUS

FOR THE

USE OF AMATEURS

CHARLES HOLTZAPFFEL'S

PRINTING APPARATUS

FOR THE

Use of Amateurs

REPRINTED FROM THE THIRD
GREATLY ENLARGED EDITION OF 1846
AND EDITED BY

JAMES MOSLEY
& DAVID CHAMBERS

PRIVATE LIBRARIES
ASSOCIATION
1971

Reprinted by photo-litho from the
third edition of 1846

Published by the
Private Libraries Association
41 Cuckoo Hill Road, Pinner,
Middlesex

ISBN 900002 60 3

Printed in Great Britain at the
Gresham Press
by Unwin Brothers Limited
Woking & London

TABLE OF CONTENTS

LIST OF ILLUSTRATIONS

INTRODUCTION

The ancestor of the 'common press' used by printers until the mechanical improvements of the nineteenth century was the screw press used in antiquity for extracting oil and juice from olives and grapes and in the later middle ages for removing the surplus water from newly-made paper. Its introduction can be dated with some precision to the first century A.D. Pliny, in his *Natural History*, speaks of it as having been introduced 'within the last twenty years'. Its predecessor, which was in common use in the countries of the Mediterranean from 1000 B.C., was a beam-press, comprising a lever of the second order, that is, one in which the 'weight' is situated between the fulcrum and the power.[1]

THE BELLOWS PRESS

It has never been seriously doubted that the screw press, in a highly developed form, was used by even the earliest printers for the printing of books. But there must always have been a potential use for a smaller, simpler and cheaper press for the printing of labels and other small ephemera without occupying the screw press which was better employed on more important work. The 'bellows press', which is claimed in the preface to Holtzapffel's handbook as the ancestor of his Parlour Printing Press, is a reversion to the principle of the simple beam-press. It is ignored in the manuals of Moxon and his successors, and we do not know for certain when it was accepted, however grudgingly, into the printing office. The earliest reference to it by name[2] that we have found is in 1781 in the sale catalogue of the property of an unnamed printer, now known to be Edward Rowe Mores,

1. Charles Singer (etc.), eds, *A history of Technology*, (Oxford (1956) 1957), ii, Chapter 4: 'Food & Drink' by R. J. Forbes.
2. E. Rowe Mores, *Dissertation*, ed. H. Carter and C. Ricks (Oxford, 1961), p. 128. Two presses constructed on this principle, one of them said to date from the eighteenth century, are displayed in the Musée du Conservatoire des Arts et Métiers, Paris, nos. 510 and 8404. Another, of uncertain age, but possibly also eighteenth century, may be seen at Newarke House, Leicester.

antiquary, amateur printer and the first historian of English type-founding. It was sold for 5*s* 6*d*.

The bellows press in its elementary form is clearly a weak contrivance, but it has the great advantage of rapid operation and extreme simplicity and cheapness of construction. The principle of a hinged platen was one that was to be pursued with varying degrees of success in its various forms. In the 1790s it was used with great ingenuity by Philippe-Denis Pierres, *Imprimeur du Roi*, who combined it with a lever and crank which operated on the platen by means of hooks at each side.[1] A little later, J. Stafford, a printer and printers' engineer of Bingham in Nottinghamshire, invented a portable iron printing press, with a hinged platen operated upon by a lever, a screw, and a hand-lever. Hansard thought that presses of this construction would be unlikely to be made in any size larger than foolscap, and criticised the press for the dangerous concussion likely to result from a too hasty throwing open of the very heavy platen and its counter-weights.[2] Hansard also described Treadwell's press (patented in 1820, No. 4433), a very large construction having a hinged platen and a series of levers moved by a treadle to effect the impression.

These and similar experiments with the bellows principle led eventually to the jobbing platens perfected by the Americans and still in wide use today.[3] But it was not until the 1870s that vertical hand-platens were made—the Kelsey, the Model (and, much later, the Adana)—of a simplicity and size suitable for amateur work.

Thirty or forty years before, in designing his Parlour Printing Press, Edward Cowper went back to the primitive form (albeit with a sophisticated lever) of the simple flatbed bellows press. This had its ancestry in the modest presses of Moore and Pine whose purpose is clearly shown in a handwritten copy of a trade

1. David Chambers, 'An Improved Printing Press by Philippe-Denis Pierres', *Journal of the Printing Historical Society*, 3 (1967), pp. 82–92.
2. T. C. Hansard, *Typographia* (London, 1825), p. 657 and plate opp. p. 696. The only known surviving example, formerly in the Brighton Museum and Art Gallery, is now in the collection of the St Bride Printing Library. It stands, with its wooden frame, 42 inches from the ground, has a platen 12¼ × 7⅞ inches, and with its quadruple counterweights probably weighs about a hundredweight.
3. A detailed account of machines of this sort is given by Ralph Green, *A History of the Platen Jobber* (Chicago, 1953).

notice (written in a copy of Smith's *Printer's Grammar*, London, 1787, now in the possession of James Mosley) offering card presses ranging from $3\frac{1}{4}$ by $4\frac{3}{4}$ inches to 14 by 17.

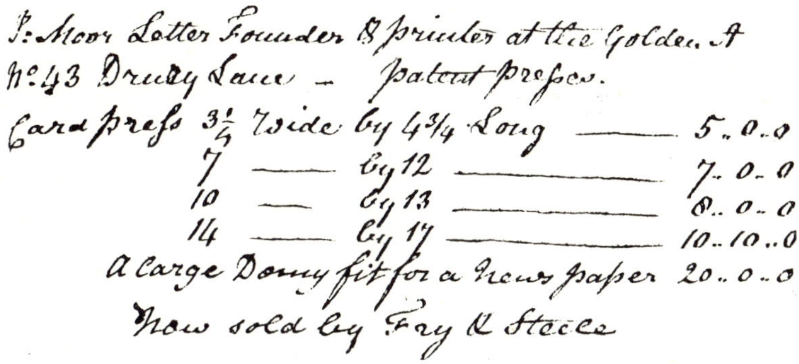
J: Moor Letter Founder & printer at the Golden A
No 43 Drury Lane – patent presses.

Card press	3¼ wide	by 4¾ Long	——	5..0..0
	7 ——	by 12	——	7..0..0
	10 ——	by 13	——	8..0..0
	14 ——	by 17	——	10..10..0
a Large Demy fit for a news paper				20..0..0

Now sold by Fry & Steele

Fig. 1. MS. Price List of Card Presses in a copy of J. Smith, Printer's Grammar, *1787, now in the possession of James Mosley.* (*Reduced*)

These presses were patented in 1771 (patent No. 999). Their power was obtained by the use of a long lever, from which the platen was loosely suspended, plainly to avoid the disadvantage inherent in presses of the bellows kind, where variation in the packing of the platen produces excessive pressure at one end or

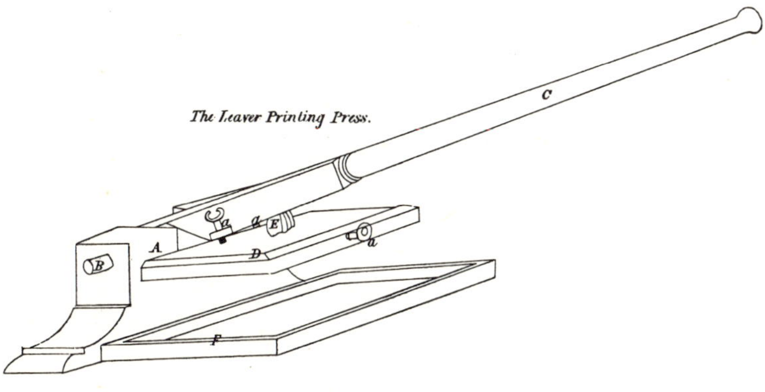

Fig. 2. Moore and Pine's Lever Printing Press, from the patent specification, no. 999, 6 November 1771, printed London, 1856. (*Reduced*)

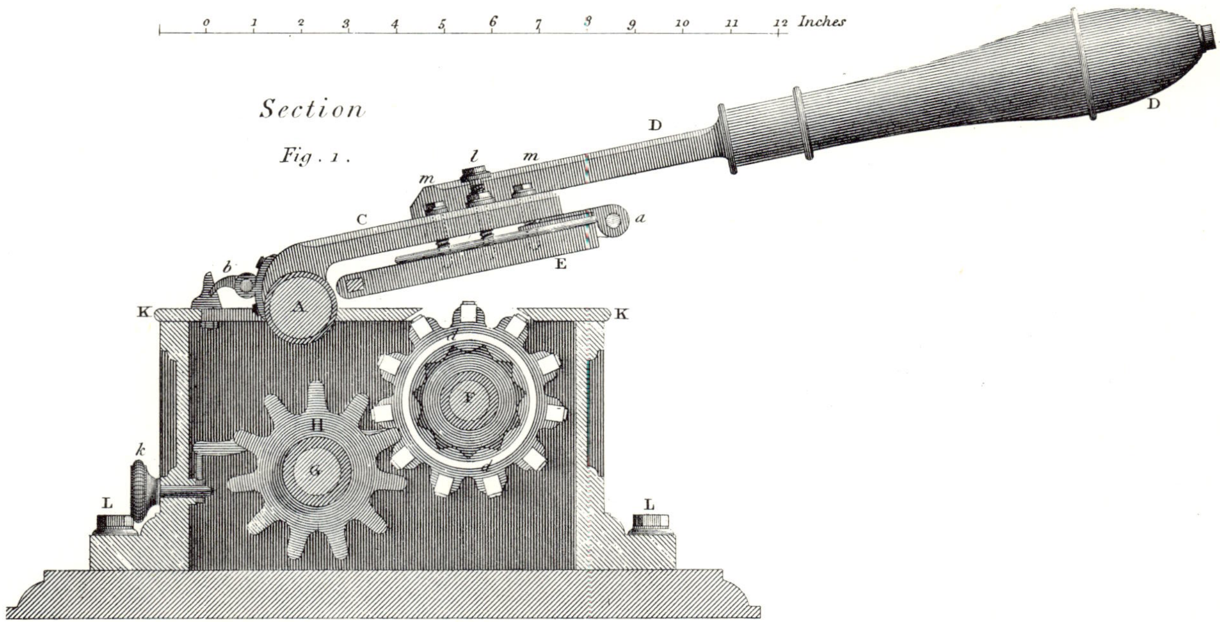

Fig. 3. Bramah's Bank Note Numbering Machine, from Abraham Rees, Cyclopaedia, *1820.*

other of the forme. They seem to have had little success, for nothing like them is mentioned in any of the handbooks of the period. In any case, in the larger sizes the use of the long lever would only be effective if the sub-structure were made so substantial as to lose the press its advantages of simplicity, cheapness, and lightness of construction. And at £5, the smallest of the presses, capable only of printing a small trade card, were very expensive.[1]

The bellows principle was used to good effect by Joseph Bramah in his bank note numbering machine (patented in 1806, No. 2977) by means of which the laborious work of many clerks was superseded by the rapid impression of a single line of figures, using type cast for the purpose by the Caslon Letter Foundry. Holtzapffel, on page six of his handbook, refers to Bramah's press but remarks that a press of this elementary sort (where the impression lever is simply an extension of the platen) is really only effective when printing a very small area of type.

Cowper's innovation, turning his attention from the massive complexities of his steam-driven machines, was to apply to the bellows press the toggle or knee joint which had already revolutionised the power of the conventional hand-press.[2] By adding a

1. John Wilkes, in 1763, paid £14 for his common press.
2. It may well be that Cowper was following an earlier precedent here. There is, in the Salford Art Gallery and Museum, a wooden bellows press with a lever pivoted at its open end, which looks as if it dates from the eighteenth century. (Perhaps one of Sutter's Portables? See page xv.)

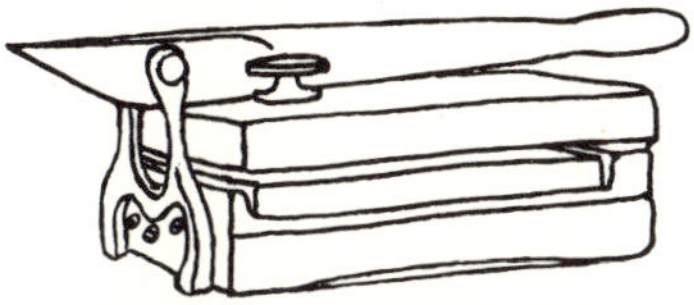

The press is made of pine, apart from some oak which is used for the handle, chase, and a quarter-inch lining to the bed. The platen area is $9\frac{1}{4}$ by 6 inches, the bed 10 by 6 inches, the lever is $15\frac{1}{2}$ inches long, but only 13 inches from the pivot. The chase has inside dimensions of $6\frac{1}{2}$ by $4\frac{3}{4}$ inches. The bed, including the oak lining, is $1\frac{1}{2}$ inches thick; the platen is $1\frac{1}{8}$ inches thick. The press bears a striking resemblance to Cowper's. The lever, too, is of the second order, and increases in power, like Cowper's, as it descends. It is attached to a substantial iron stirrup plate screwed to the end of the bed, and bears upon the edge of the platen until this is forced into a horizontal position; it is curved beyond the pivot so as to clear the platen when this is opened after the lever is raised. We are obliged to Mr Derek Nuttall for drawing our attention to the press while this introduction was in proof, and to Miss Evelyn Vigeon, Keeper of Folk Life at the Museum, for supplying details and photographs (from which the accompanying sketch was taken).

simple toggle to the open end of the bellows he was able to design a press which combined the simplicity and cheapness of the bellows press with the power necessary to print a large forme. The action of his lever was such that its power was greatest at the end of its downward sweep, and the rest of the press was made robust enough to withstand the considerable pressure that could be exerted. The hinges, pressure plate, lever, toggle and eye were all of at least $\frac{3}{8}$-inch iron. Right-angled hinges were employed, let into the platen and bed and screwed (with $1\frac{1}{4}$-inch, size 14 screws) from above and below as well as into the ends. This was a far stronger method of fixing them, and also had the advantage of allowing exact adjustments to be made in the opening of the press by means of packing placed under or over the right-angle pieces. In the smaller presses the platen and bed were made of two-inch mahogany, and in the foolscap-folio and half-sheet demy the bed was strengthened by a half-inch cast-iron plate. The largest size, not mentioned in the 1846 price list, but listed in a later four-page leaflet (see fig. 8), was made entirely of iron. The foolscap-folio could be had either with or without a counter-poise to the platen; the larger presses were always supplied with this aid to easier working.

We do not know exactly when the first Parlour Presses were made. The earliest references to them that we have so far traced are in the preface to G. Levinge, *Traveller in the East*, 1846, in which the author tells us he bought a press in the summer of 1839, and the first edition of Holtzapffel's manual itself, published in 1839. Holtzapffel's General Catalogue of 1834 includes tools for printers on its title-page, and lists, for the first time, 'Printing Presses on the Stanhope and other principles, of small sizes'. Parlour Presses are not mentioned, and perhaps we may assume they were not manufactured until about 1839 when the manual that was written to instruct amateurs in their use was first published. Some thirty years after the introduction of the Parlour Printing Press, Jabez Francis of Rochford, Essex, marketed a series of devices which superficially resemble it. They were in fact quite different in principle.[1] The lever was not actually applied to the platen but raised the bed from below by acting upon two interlocking U-shaped bars. Moreover, these

1. An example, with a $12 \times 9\frac{1}{2}$ inch bed, is in the possession of the St Bride Printing Library.

interlocked U-pieces raised the bed, inside the frame to which the platen was hinged, parallel to the platen and not in a bellows movement at all. The construction was all of iron and the weight of the medium size, to print an area of $9\frac{1}{2}$ by 8 inches, was fifty pounds. A writer in the *British and Colonial Printer* for 4 July 1902, talking of the craze for amateur printing which swept the country between 1875 and 1885, described them as great favourites with those who aspired to good work. The frontispiece to the second edition of Francis's handbook, *Printing at Home*, published about 1873, is the only attempt known to us to represent parlour printing at its most idyllic, with fond parents admiring the productions of their busy children while the kitten unheeded chases a ball of wool beneath the heavy mahogany table. (See fig. 4.)

There were other variations on the theme of the flatbed bellows press before the end of the century, by which time the frenzy for amateur printing had to a large extent died away. The later revival, using vertical platens, was based on the idea of printing as a paying hobby rather than an elevating family pursuit. For those still interested in bookwork the Adana flatbed has taken the place of the Parlour Press which, in its early, wooden models it greatly resembled. But it is worth noting that the lever of the automatically inked, all metal, flatbed Adana of today, though quicker in its application, nevertheless lacks the advantage of increasing power that was so ingenious a feature of its Victorian ancestor.

THE PARLOUR PRESS IN USE

Cowper's was the first really small, simple, yet effective press for amateur use, and it is not easy to decide whether its appearance was the cause or the effect of the popularity of printing as a Victorian middle class pastime.[1]

In the eighteenth century, private printing had chiefly been the amusement of those with the means to employ their own compositor and pressman working with the conventional equipment of the trade. Horace Walpole set a pattern at Strawberry Hill

1. For a more general survey of the private printers flourishing at this time the reader may be referred to Roderick Cave's history, *The Private Press* (London, 1971). A fuller account of the mechanical antecedents and descendants of the Parlour Printing Press will be found in James Mosley, 'The Press in the Parlour', *The Black Art*, Vol. 2, No. 1 (1963).

Fig. 4. Frontispiece to J. Francis, Printing at Home, [*1873?*].
(*St Bride Printing Library*)

which was followed in their various ways by John Wilkes in Great George Street, Thomas Johnes at Hafod, and Sir Egerton Brydges at Lee Priory. Those without comparable means or ambition made do with other expedients. Edward Rowe Mores's often-quoted account of the work of 'Mr Stevens' is the most charming example because it conveys the half-intoxicated delight with the potential power of printing which lies behind so many of even the most apparently innocent and humble private presses.

> *Mr Stevens* was a gentleman of a typographical turn, but no great adept. he purchased some letter at *The Hague*, and when he came home he printed for his recreation, he used wooden chases nailed upon planks: no composing stick: no head-sticks, foot-sticks, side-sticks, gutter-sticks, quoins, or other furniture, but nails only with which he pegged his matter together: his balls were a bunch of waste paper: his tympans and frisket a dirty handkerchief: his press for small work the ball of his thumb; for larger a rolling-pin and old rags. he was an antient bachelor of odd humour and of *Dutch* taste, in his garb and gesture antique indeed, and the furniture of his house was of the reign of *Qu. Eliz.* the work in which he delighted was below the degree of *Drops* or *Patters* or *Chaunts* or *Runs*. he devised and printed *title-pages* of strange and ludicrous books *speedily to be published* which were never to be published, nor indeed had any existence; and these title pages he dabbed up in the cool of the evening at the corners of the public streets to stir up the expectation of those who stopped there.—this was *his* amusement, and harmless enough.[1]

'For work of this bulk wooden chases may suffice,' says Mores. 'Sutter's portables are little more.' 'Sutter's portables', advertised in 1769 as a means whereby 'the young Nobility and Gentry of both sexes' might be made acquainted with the beauties of the Poets and Authors, may well have been a variant of the bellows press which Mores himself owned.[2] In 1776 Samuel Johnson and James Boswell visited the museum of the Litchfield apothecary Richard Green, who had 'all the articles accurately arranged with their names upon labels, printed at his own little press'[3]

1. E. Rowe Mores, *Dissertation*, ed Carter and Ricks (Oxford, 1961), p. 66.
2. J. Sutter was an engraver and printer in St Martin's Churchyard, St Martin's Lane. In the *Reading Mercury and Oxford Gazette* for 30 October 1769 he advertised, 'to Noblemen, Gentlemen and Ladies curious in Printing', his 'Portable Printing Presses, on a new construction; by which two folio pages may be printed, more or less as required, with the greatest expedition and exactness.' Type and other materials were supplied, 'with proper instructions and examples, how to compose the Types, and cast off the Copies'. He also advertised 'Small Printing Presses, the size of a pocket volume', for printing cards. We are grateful to Mr James Moran for this reference.
3. James Boswell, *Life of Johnson*, 23 March 1776.

which must have been some such device. Indeed they may have been something of an English speciality, for the Breitkopf typefoundry offered for sale a 'Little English pocket printing office for marking washing',[1] and the Italian poet and dramatist Vittorio Alfieri printed his sonnets for his own amusement with a 'little printing office' (*stamperiuola*), which would serve for exactly fourteen lines and no more, while recovering from an illness shortly after returning from a visit to England in 1786.[2] In 1815 Sir Alexander Boswell himself began the work of the Auchinleck Press with a miniature press made by Ruthven, which was soon replaced by a full-sized Ruthven press, worked by a professional pressman.[3]

None of these miniature presses had been really suitable for uninstructed domestic use. When a leisured middle class came into being in the early nineteenth century ready to take up this elevating hobby, it was Holtzapffel who supplied them with the means to do so. His small, easily operated, moderately priced presses,[4] together with (what was just as important a matter for the convenience of amateurs) a ready supply of small quantities of type and other equipment, gave parlour printing its first wide popularity. The varied uses to which his press might be put are suggested in his later Price Lists (see fig. 8):

> Companies, institutions, and individuals, have found it convenient for circular letters, invoices, and papers, subservient to the despatch and methodical arrangement of business; naturalists and travellers for short memoirs of scientific researches, or labels for specimens; the friends of education for disseminating original and other papers; wood-engravers, for examining the progress of their blocks; practical printers, for proofs of title-pages, stereotype plates, or cards; and nearly every different pursuit will suggest some new application of this little Press.

Most of the Victorian amateur printing so produced has perished, and perhaps this is not too serious a loss. Some idea

1. 'Kleine englische Taschen-Druckerey, zum Wäsche zeichnen', *Preis-Courant von Schriften, aus der Breitkopfischen Schriftgieserey* (Leipzig, 1787). Perhaps made by Sutter, who advertised 'stamps' for marking linen, as well as his portable presses.
2. *Lettere edite ed inedite di Vittorio Alfieri a cura di Giuseppe Mazzatinti* (Turin, 1890), p. 149.
3. Sir Alexander Boswell to Dibdin, 5 May 1817, *Bibliographical Decameron* (London, 1817), iii, p. 453.
4. *Moderately* priced, but the smallest outfit cost £5 6*s* and one might compare this with the expense to the Westmeath Protestant Society (whose early reports were printed with such an outfit) for the diet, washing and lodging of an orphan, which amounted to just £3 10*s* per annum.

of its extent and scope can be gained from a careful study of Madan's bibliography of the Daniel Press[1] which lists innumerable pieces of ephemera, much of which can be seen in the collection preserved at the Bodleian Library. Notices, announcements of garden fêtes, tickets, prayers, birthday letters, subscription requests, book-plates, titles, labels, then as now were the staple work for the spare-time printer. The more ambitious printed small books, a number of which have survived, and of these a few can be shown to have been printed on Parlour Presses made by Holtzapffel.

One body of such work was produced by Godfrey Levinge,[2] born in 1796, who travelled widely in the east before settling in 1840 at Cullean, near Mullingar in Ireland. He wrote an account of his travels which he determined to print himself, and a first volume of which he completed despite the fact that much of it was absorbed into one of Murray's hand-books[3] while he was still printing in Ireland. *The Traveller in the East* . . ., Printed by the Author, Curzon Street, Mayfair, 1839, was published in 1846; the projected second volume was never printed owing to the famine which struck Ireland in 1847. An account of the circumstances of his setting up his press is given in the preface to the first volume:

'In the Summer of 1839 my labours as author having ended; I was in London without occupation, and accidentally seeing at Mr. HOLTZAPFFEL's establishment Cooper's parlour printing press, I determined on learning the art of printing; I purchased one of them, had a printer to instruct me, and my first typografical production is the introduction to this volume; as I acquired knowledge of my new trade I became dissatisfied with my type, and changed it twice before I got the font with which I work at present; this will account for three different kinds of printing appearing in the following pages. . . . In May, 1840 an

1. Falconer Madan, *The Daniel Press 1845–1919* (Oxford, 1921).
2. We are much indebted to Mr N. W. English for very kindly bringing Levinge's work to our notice, Miss Marion Keaney, of the Longford-Westmeath County Library, for supplying further biographical information and to the Library for allowing us to reproduce the page from the Westmeath Protestant Orphan Society reports in its possession.
3. *A Hand-Book for Travellers in the Ionian Islands, Greece, Turkey, Asia Minor, and Constantinople; being a Guide to the principal Routes in those Countries, including a description of Malta; with Maxims and Hints for Travellers in the East* (London 1840.)

ABSTRACT OF THE CASH ACCOUNT TO THE FIRST OF MAY, 1841.

Received.

	£	s.	d.
Amount of Donations for 1840, ...	63	11	0
Annual Subscriptions for ditto, ...	137	11	6
	£201	2	6

Expended.

	£	s.	d.
Postage Stamps on Circular Letters, ...	0	12	0
Paper for Printing 500 Copies of 'Rules and Regulations',	0	16	0
Do. for Circulars, ...	0	1	6
Treasurer and Secretary's Books, ...	0	8	6
To Mr. Purdue, for Printing a various stock of Forms of Application for Orphans, Nurses, &c., &c., that will last for several years; and for Advertisements of the first year of the society, ...	15	0	0
To Clothing for Twenty Orphans, advanced on their election	5	0	0
Rent of Hotel for Meetings, and attendance, ..	0	18	6
Nurses' Salaries paid up to 1st May, 1840, ...	30	12	1
Apprentice Fees lodged for Twenty Orphans, ...	100	0	0
Lodged in Castlepollard Bank to the credit of the Society,	47	13	11
	£201	2	6

GODFREY LEVINGE, *Vice-Treasurer*.

WE, the undersigned, having examined the above statement, do find it correct, and that a Balance of £47 13s. 11d. is lodged in the Castlepollard Saving's Bank, to the credit of the Protestant Orphan Society.

ALEXANDER NIXON,
CHARLES BAYLY, } *Auditors*.

important alteration took place in my manner of life. I became the purchaser of an Estate, established myself permanently as a resident in the country and undertook the management of 200 Irish acres of land, which, in addition to the public duties imposed by a country residence has allowed but little time for printing, except on rainy days and long winter nights, so that the printing of this volume has extended over a period of nearly seven years.' Issued in a pleasing trade binding of purple cloth (with a vignette of a traveller, camels, palms, and pyramids, in gold on the front) *The Traveller in the East* is a substantial volume running to 331 demy octavo pages. It is set in a modern face pica type to a measure of 24½ picas, three point leaded, 31 lines to a page. Anyone who has attempted the printing of such a book, on a hand-press, a page—or two—at a time, will come to the conclusion that there must have been many rainy days in Cullean between 1840 and 1846.

During this same period Levinge also found time to print the first six annual reports, for the years 1841–6, of the Westmeath Protestant Orphan Society, extensive affairs usually running to thirty or forty pages and including accounts, subscription forms, tabular lists of orphans, lists of donations and lengthy reports of good works.

The typographic style of book and reports is neat and unostentatious—typical of commercial work of the period; the press-work is reasonable, though the inking tends to be uneven. Unfortunately there is no indication of the number of copies printed but an edition of up to 200 might be deduced from the list of 160 subscribers to the Protestant Orphan Fund in the *First Report.*

The next work which we know to have been printed on a Parlour Press is a slim quarto of 52 pages, dated 1840. *Attempts in the Art and Mystery of Printing by an Amateur*, bears the imprint of Nathaniel Brown Engleheart, Blackheath. The matter, like that of much English private printing of the late eighteenth and early nineteenth centuries, is antiquarian: *The Chronicle of the Kings of England* and *The Memoir of Guy, Earl of Warwick*, transcribed by the printer 'from a Monkish Manuscript'. A disarming note asks the reader to bear in mind 'that this little

Opposite: Fig. 5. Page from the First Annual Report, *Westmeath Protestant Orphan Society, 1841.* (*Longford-Westmeath Library*)

Volume is ushered into existence, not as a Specimen of Antiquarian Research; but as one of Amateur Printing'. The first leaf of the book, a kind of prefatory half-title which we reproduce here (fig. 6), shows that Engleheart handled the types and ornaments supplied by Holtzapffel with great charm and skill. He also owned one of Holtzapffel's lathes, with which he engraved delicate spiral patterns on wood and metal by means of an eccentric chuck, His handbook of the technique, *A Concise Treatise on Eccentric Turning . . . by an Amateur*, published in 1852, expresses his glowing appreciation of the tools which they supplied. It is commercially printed, but contains impressions of fifty of his patterns printed 'with his own hands' on the Parlour Printing Press.[1]

The only other body of work that we know to have been printed on a Parlour Press is that done for Major-General (then Lieutenant-Colonel) Sir William Sleeman when Resident at the Court of Lucknow. His *Diary of a Tour through Oude*, 1852, was an account written for submission to the Government of India 'of the real state of the country, condition and feeling of the people of all classes, and character of the government under which they at present live. . . .' He had 18 copies printed at his own expense 'in a small parlour press' which he purchased, with type, for the purpose.[2] The report largely influenced Lord Dalhousie in his resolve to annex the kingdom, although this course was opposed by Sleeman who thought that reforms could be accomplished under native rule. The book makes two volumes of 309 and 302 well-filled pages. The presswork is on the whole good and

1. We owe our knowledge of Engleheart's *Attempts in the Art and Mystery of Printing* (and his printing of the plates in the *Concise Treatise on Eccentric Turning*) to Mr J. Leggat of Stirling. We are most grateful to him for allowing us to examine his copy and to reproduce a page from it.

2. Letter from Sleeman, 2 January 1853, printed as appendix to the reprint of the *Diary* under the title *A Journey through the Kingdom of Oude, in 1849–1850* (London, 1858), vol. 2, p. 393. Some half dozen copies of the *Diary* were distributed to government officials, five went to children, and one to a brother. The government copies were 'coarsely bound' as Sleeman could find 'no good binder' in Lucknow. The copy in the Bodleian (for knowledge of which we are indebted to Michael Turner) was perhaps his own copy (it bears the signature A. J. Sleeman, presumably that of his wife, Amélie Josephine); it has numerous added corrections and notes, both printed and in manuscript, and was well bound by Martin of Calcutta.

Opposite: Fig. 6. Half-title of Attempts in the Art and Mystery of Printing. (*Mr J. Leggat*)

Specimens

of

AMATEUR PRINTING.

THESE

SPECIMENS OF AMATEUR PRINTING

Were effected by means of the ingenious little Press

Invented by

Mr. COWPER.

and manufactured by

Messrs. HOLTZAPFFEL & COMPANY.

Of No. 64. Charing Cross & No. 127. Long Acre.

Mr. Cowper's Talents have been successfully employed, in designing both extremes of printing Machines ; from the colossal TIMES Newspaper Machine, - which prints 4200 Sheets per hour, or 70 Papers in a minute, measuring 48 inches by 36 inches; to the present miniature and fire-side Apparatus. (See Book of Instructions by H. and Co.)

remarkably even except for the preface and tipped-in errata sheets which are very heavily impressed. There are 31 or 32 lines, plus headline, to a page; the type is a modern face long primer, heavily leaded, set to a measure of 24 picas, making a type area of about 7¾ by 4 inches. On one page a column of notes, set in unleaded brevier, 10 picas wide, was printed at a second impression in the outer margin. Elsewhere in the copy of the *Diary* in the Bodleian similar columns of notes, plainly afterthoughts (and not included in the 1858 reprint), have been pasted in, asterisks being added by hand to the text. These extra notes are not present in the copy of volume 1 of the *Diary* in the India Office Library, London; this copy also lacks the errata and the preface.

Tipped into the front of the Bodleian copy (but not into that in the India Office) is a four-page untitled leaflet, consisting of notes on the Dynasty of Oude. It is corrected in what appears to be the author's hand (there are further manuscript notes at the end), is very heavily impressed, and is presumably a proof. The type area is almost exactly the same as that of the book; the pages are numbered 1–4 and are graced with a signature, 1 a, at the foot of page 1.

Two other sets of proofs (of which there are reproductions in the India Office Library) are now in the Ames Library of South Asia, Minneapolis, Minnesota. Both are set in a style closely resembling that of the *Diary*. One is a 100-page *History of the Reigning Family of Oude*, the other, entitled *The Byza Bae*, is a 32-page account of the troubled regency of that lady at a time when Sleeman was political resident in Gwalior.

There must have been other books printed by amateurs on presses made by Holtzapffel, though evidence is so far lacking to show which. In 1846 for instance the young Henry Daniel, at the age of nine or ten, was given 'a printing press with all its appendages,' including a box and types.[1] Madan refers to this as a toy, but it may yet have been a Parlour Press, though it

1. Printed letter of thanks from Henry Daniel to his father, 1846, listed by Madan in *The Daniel Press 1845–1919* (Oxford, 1921), pp. 70–1.

Opposite: Fig. 7. Title page of Diary of a Tour through Oude, *1852.* (*Bodleian Library*)

DIARY

OF A

TOUR THROUGH OUDE,

IN

DECEMBER 1849, & JANUARY & FEBRUARY, 1850.

BY

THE RESIDENT

Lieutenant-Colonel W. H. Sleeman.

VOLUME I.

PRINTED AT LUCKNOW IN A PARLOUR PRESS.

1852.

would have been an expensive enough present for a country curate to give a son of Henry's age. Gaetano Polidori was another who set up a private press at home and who may conceivably have used a Parlour Press. No amateur, he was head of a firm of printers and publishers in Little Cockspur Street, north of Trafalgar Square, and is most likely to have seen Holtzapffel's presses at 64 Charing Cross, facing the south side of the Square. His earliest recorded books were a three volume set of Milton in Italian, issued in 1840, which would have been soon after the presses were first manufactured. But apart from Levinge, Englehart, and Sleeman the names of users of Cowper's invention are so far a matter of speculation and the editors would be glad to hear of any work the printing of which can be definitely credited to Holtzapffel's Parlour Printing Press.

THE MANUAL

The only manual for amateur printers earlier than Holtzapffel's known to survive is a sixteen-page unillustrated pamphlet by F. C. Ritter, *Nachricht von zweyerley Arten neu erfundenen Taschen-Buchdruckereyen sowol mit als auch ohne Buchdrucker-Presse nebst hinlänglichen Unterricht von deren Gebrauch,*[1] published in Hamburg in 1772, and printed on one of the presses it describes. English speaking amateurs will hardly have come across it, and until Holtzapffel published his description they would have had to rely on the works of Moxon and his successors intended for the guidance of professional printers.

The first edition of the handbook which is now reprinted was presumably the octavo volume *Complete Instructions for the management of Mr Cowper's parlour printing Press* (London, 1839) which is listed in the Catalogue of the St Bride Printing Library issued in 1919. Unfortunately this does not appear to have been seen since it was catalogued and we have been unable to locate another copy, nor for that matter, to discover any reference anywhere to a second edition. Copies of the third edition, 'greatly enlarged', are relatively easy to locate and have possibly survived because more substantial than their predecessors. No later

1. 'Account of two kinds of newly invented pocket printing offices both with and without a printing press, together with instructions for their use'. The only copy known is in the Deutsches Buch- und Schriftmuseum in Leipzig.

editions have been found, but there are variants in the gold and blind blocking of the boards, and a copy now in the possession of James Mosley has a price list, printed between 1873 and 1897, pasted into the back in place of the original price list on pages 78 and 79. (See fig. 8.) It may be presumed therefore that the third edition remained in print for at least thirty years, and possibly for much longer, and that all the sheets were not bound up at once.[1]

The later price list was also issued separately, for its first page (pasted down to page 78 in James Mosley's copy of the manual) has printed on it: PRICE LIST / PRINTING APPARATUS FOR THE USE OF AMATEURS / MANUFACTURED & SOLD BY / HOLTZAPFFEL & CO., / Engine, Lathe, and Tool Manufacturers / 64, CHARING CROSS, & 127, LONG ACRE, LONDON. That it was printed in 1873 or later can be shown by its having been printed by Bradbury, Agnew & Co, whose name in this form first appears in the Post Office London Directory in 1874; it must have been printed before 1898 because in that year Holtzapffel's address is shown as Charing Cross and Sharples Hall Street. Prices hardly changed in the years between 1846 and its publication: the best apparatus complete now cost £8 instead of £7 2*s*; a duplicate set of 2500 types still cost £1 12*s*; the folio foolscap outfit, in the less complete form but with iron bed and counterpoise, cost £22 instead of £17 14*s* 6*d*; and in its more complete form £32 instead of £27 13*s* 6*d*. These presses were no longer available separately, but had to be bought as part of complete outfits. The rollers supplied were vulcanised instead of composition. The folio demy press (half-sheet demy),

1. Three variant bindings have been noticed in the four copies that have been studied (belonging to *a*. Berthold Wolpe, David Chambers; *b*. St Bride Printing Library; *c*. James Mosley). The earliest seems to be (a) that in green cloth blind-blocked with scrolls in the corners, with the title blocked in a condensed modern face, and the words PRINTING APPARATUS measuring $2\frac{11}{16}$ inches. That which is perhaps next in date (b) is in a darker green cloth, blind-blocked with leaf scrolls in the corners, with the title in Caslon Old Face apart from the words FOR THE which are set in a condensed modern, and with the words PRINTING APPARATUS measuring $3\frac{7}{8}$ inches. The latest issue (c, copy with price list of 1873/1897) is in blue-green cloth sharply blind-blocked with crosses inside circles in the corners, with the title in a large condensed modern with a full point at the end, and the words PRINTING APPARATUS measuring $3\frac{15}{16}$ inches. Other variants probably exist.

Overleaf: Fig. 8. Inner pages from Holtzapffel's Price List of Printing Apparatus, *issued between 1873 and 1897.* (*James Mosley*)

COWPER'S PARLOUR PRINTING PRESS.

MADE ONLY BY HOLTZAPFFEL & CO.

64, Charing Cross, and 127, Long Acre, London.

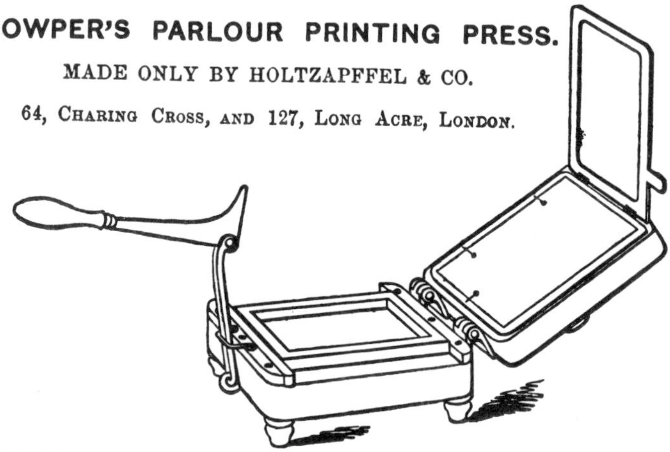

This little Printing Press is made of mahogany, and stands in the small space of 11 by 8 inches. It is capable of printing a page 7 by 6 inches, and works so easily that a child may use it on the parlour table. A small type case accompanies it, containing a fount of about 2,500 types, neatly arranged in three drawers with appropriate divisions; a fourth drawer serves for the furniture, inking tablet, &c.; and to these are added the necessary tools, so as to render the whole complete. Should it be required, the type case will contain a duplicate supply of type in addition to that usually furnished, and which doubles the efficiency of the apparatus at a slight additional cost.

The above apparatus is well adapted for the amusement and education of youth, and also to various applications of the inestimable typographic art to the common concerns of mankind.

For example.—Companies, institutions, and individuals, have found it convenient for circular letters, invoices, and papers, subservient to the dispatch and methodical arrangement of business; naturalists and travellers for short memoirs of scientific researches, or labels for specimens; the friends of education for disseminating original and other papers; wood-engravers for examining the progress of their blocks; practical printers, for proofs of title-pages, stereotype plates, or cards; and nearly every different pursuit will suggest some new application of this little Press.

LIST OF PRICES.

SECTION I.—Cowper's Parlour Presses and Apparatus.

COWPER'S PARLOUR PRINTING PRESS, with a galley-chase, a box of ink, a vulcanized inking roller, and a distributing tray.

SMALL DEAL TYPE CASE, painted, with four drawers; three of them partitioned to contain an assortment of about 2,500 types, and a proportionate supply of leads and brass rule; the fourth drawer contains reglet, furniture, side and foot sticks, quoins, &c.

SET OF EXTRAS—comprising transfer composing stick, bodkin, forceps, mallet, shooting-stick, planer, brush, and turpentine for cleaning the type, two quires of demy printing paper, cut into suitable sizes for the press, and one pair of damping slates.

GALLEY CHASE seven inches square inside.

Total charge for the Plain Parlour Press and Apparatus complete, £6.

COWPER'S PARLOUR PRINTING PRESS, japanned, and finished in the best manner, and fitted with a drawer, in other respects as above.

SMALL MAHOGANY TYPE CASE, with brass lock and handles, in other respects as above.

SET OF EXTRAS, comprising transfer composing stick, &c., as above.

GALLEY CHASE seven inches square inside.

Total charge for the Best Parlour Press and Apparatus complete, £8.

DUPLICATE SET OF 2,500 TYPES, and which may be contained in either of the above cases, £1 12*s*.

SECTION II.—FOLIO FOOLSCAP PRESSES AND APPARATUS.

FOLIO FOOLSCAP PRINTING PRESS, on the principle of Cowper's Parlour Press, suitable to printing the half sheet of foolscap, or the quarto sheet of imperial. With an iron bed half an inch thick, planed quite level and true, to increase the permanent accuracy of the Foolscap Press, and an iron counterpoise to facilitate the working of the same. Measurement of the bed, 15 by 10 inches. The press varnished and japanned, complete, with two iron chases, register points, &c.

LARGE DEAL TYPE CASE, with six drawers, and measuring externally 24 inches by 18, and 11 inches high, with iron handles, lock and key. Four of the drawers are partitioned after the Printer's method for holding 9000 types of the following varieties.—GREAT PRIMER ROMAN, Specimen No. 9—viz., capitals, figures, points, spaces and quadrats. PICA ROMAN, No. 13; large and small capitals, lower case (small letters), with accented vowels for printing the foreign languages, figures, points, spaces, quadrats, and space-line leads, complete. BOURGEOIS ROMAN, No. 17; capitals, figures, points, spaces and quadrats. BOURGEOIS ANTIQUE, No. 23; capitals, figures, points, spaces and quadrats. Two of the drawers contain space-line leads, furniture, side and footsticks, quoins, and reglet; also a mallet, shooting-stick, planer, bodkin, printer's composing stick 9½ inches long, brush for cleaning the type, a pair of thick damping slates, &c., all proportioned to the size of the Foolscap Press. Six-inch vulcanized inking roller, in frame and case, and large box of superfine printing ink.

Total charge for the Folio Foolscap Press and Apparatus in the less complete form, £22.

FOLIO FOOLSCAP PRINTING PRESS, exactly like the one last described.

LARGE DEAL TYPE CASE with eight drawers, similar to the case with six drawers above described, but three inches higher, and containing a considerably greater supply of each of the kinds of type specified in the foregoing description, together with the addition of Great Primer No. 9, lower case letters, Pica Italic No. 14, capitals, lower case letters, points and spaces, and Bourgeois Antique No. 23, lower case letters, making the total number of types about 17,000; together with a proportionate increase of space-line leads, furniture, &c., and with the addition of 21 pieces of brass rule of three varieties; six-inch vulcanized inking roller, in frame and case; large box of superfine printing ink; composing frame of deal, to receive the drawers of the type case when in use; inclined galley of mahogany, with moveable bottom; and four extra chases, two of them with crosses.

Total charge for the Folio Foolscap Press and Apparatus in the more complete form, £32.

LARGER PRESSES ON THE SAME PRINCIPLE.

FOLIO DEMY PRINTING PRESS, with iron bed and counterpoise; suitable to printing the half sheet of demy. Measurement of the bed 20 by 13 inches The press varnished and japanned, complete, with two chases, register points, &c., £15.

BROADSIDE FOOLSCAP PRINTING PRESS, made entirely of iron, and provided with screw adjustments to regulate the pressure as required; suitable to printing the whole sheet of foolscap. Measurement of the bed, 20 by 15 inches, complete, with four chases, register points, &c., £22.

Type cases of various sizes, and with any required selection of types. Moulds for casting rollers. Tools, &c., to order.

SECTION III.—CASES FOR ADDITIONAL TYPES.

SMALL TYPE TRAY, 10 by 6 inches, with a selection of about 600 Roman or Italic types, of small size, of either of the numbers 17 to 20, £1.

TYPE BOOK, 15 by 12 inches, with a selection of about 1,500 types, comprising eight varieties of small types for headings, cards, &c., as described in page 55 of pamphlet, £2 10*s*.

LARGE TYPE TRAY, 22 by 24 inches, partitioned after the mode of the printing office, for containing larger quantities of type of any kind, 7*s*.

HAND CHASE, in a painted case, with cushion, roller, ink, and inking tray, £1 5*s*.

PRINTING APPARATUS FOR THE USE OF AMATEURS.

A pamphlet containing full and practical instructions for the use of COWPER'S PARLOUR PRINTING PRESS, *also the description of larger presses on the same principle, and various other apparatus for the Amateur Typographer.—The pamphlet contains likewise numerous specimens of plain and ornamental types, brass rules, checks, borders, ornaments, corners, arms, &c. &c.*

Third Edition, greatly enlarged. 8*vo, cloth, price* 2*s.* 6*d.*

which was referred to on pages 7 and 8 of the manual but not in the 1846 price list, is now described and priced at £15, and a new press, the largest made, a broadside folio constructed entirely of iron, is described and priced at £22. The price of the manual, 'Third edition, greatly enlarged.' remains 2*s* 6*d*. On the last page of the leaflet the title-page to Holtzapffel's general catalogue is replaced by a detailed listing of various scales, curves, and protractors supplied by the firm—and also the Odontograph invented by the Professor Willis whose mode of printing architectural plans, etc., is described in the text of the manual.[1]

The type specimen too, pages 63–80 of the manual, was issued separately, gratis to customers. The pages were reimposed to make a twenty-page pamphlet, retaining the folios, and adding two pages, listing lathes and table cutlery, from Holtzapffel & Co.'s General Catalogue of 1844. A title-page, similar to that of the manual, was printed on the front of the buff wrappers (see fig. 9). Dated 1846, it quotes the price of the manual itself as 2*s* 6*d*, in cloth. Inside the cover the contents of the manual are listed in full.

COWPER & HOLTZAPFFEL

Both the inventor and the manufacturer of the Parlour Press are well known for other more serious activities than the supply of printing materials to amateurs.

Edward Cowper, born in 1790, patented the first perfecting printing machine in 1818, devised a system for forming composition into rollers, and in 1827 patented, with his brother-in-law Augustus Applegath, the four-cylinder machine which the latter erected for the *Times*, superseding Koenig's machine. He later

1. A note in P. E. Raynor, *Printing for Amateurs* (London) [1876], pp. 9–10, reads: '*Holtzapffel and Co.'s Cowper's Parlour Printing Press.*— This is a small mahogany press, and is made only in one size, printing a page 7 in. by 6 in. It is supplied in conjunction with an office at £6. It would probably be supplied separately, but the price is not quoted in Messrs. Holtzapffel and Co's catalogue. It is a really useful little press, but is expensive. We have one that has been in use for about eight years, and still does very good work. Address Holtzapffel and Co., 64, Charing Cross, and 127, Long-acre, London.' This was first printed in *The Bazaar*, 2 October 1875. Its reference to the availability of only one size of press conflicts inexplicably with Holtzapffel's lists of 1846 and of 1873/97.

Opposite: Fig. 9. Wrapper of the separately-issued Specimen of Types, *1846*. (*St Bride Printing Library*)

SPECIMENS OF TYPES

Extracted from the Third Edition of a Pamphlet,

ENTITLED

PRINTING APPARATUS

FOR THE

Use of Amateurs.

CONTAINING

FULL AND PRACTICAL INSTRUCTIONS

FOR THE USE OF

COWPER'S PARLOUR PRINTING PRESS.

ALSO THE DESCRIPTION OF

LARGER PRESSES ON THE SAME PRINCIPLE, AND VARIOUS OTHER APPARATUS FOR THE AMATEUR TYPOGRAPHER.

THE PAMPHLET CONTAINS LIKEWISE, NUMEROUS SPECIMENS OF PLAIN AND ORNAMENTAL TYPES, BRASS RULES, CHECKS, BORDERS, ORNAMENTS, CORNERS, ARMS, &C. &C.

AND IS

PUBLISHED BY HOLTZAPFFEL & Co.,
64, CHARING CROSS, AND 127, LONG ACRE, LONDON;
TO BE HAD ALSO OF ALL BOOKSELLERS.

1846.

Price of the Pamphlet in Cloth, Third Edition, 2s. 6d.—The Specimen of Types gratis to Customers.

went into partnership with his brother Ebenezer and supplied machines to printers throughout Britain and Europe. Towards the end of his life he was professor of manufacturing art and mechanics at King's College, London. He died in 1852.

Charles Holtzapffel was born in 1806, son of a German who had settled in London as a worker in tools and lathes. He became a skilled and educated mechanician, publishing in 1838 *A New System of Scales of Equal Parts, applicable to various purposes of engineering, architectural and general science.* His principal work, for which he is still highly regarded, is a treatise, *Turning and Mechanical Manipulation, intended as a work of general reference and practical instruction on the lathe, and the various mechanical pursuits followed by amateurs,* of which he completed the first three volumes only, before his early death in 1847, the concluding two volumes being written by his son, John Jacob Holtzapffel.

The firm of Holtzapffel was founded in 1794. Its first printed catalogue was issued around 1820; later catalogues grew to a great size and listed a large variety of tools and instruments for all sorts of tradesmen and amateurs, as well as cutlery. The title-page to the 1834 catalogue gives the address of the firm as 64 Charing Cross, late of Cockspur Street, the manufactory being behind 127–31 Long Acre. Entries in the Post Office London Directory record the later movements of the firm to Sharples Hall Street, to New Bond Street, and eventually to Eglon Mews, SW1. Amongst the specialities listed in the directory 'printing presses (for regimental & parlour use)' are listed for the first time in 1865, and, surprisingly enough, for the last time in 1927. The final entry for the firm, then simply described as 'mechanical engineers', was in 1938.

DESCRIPTION OF A PARLOUR PRINTING PRESS

The only example of a Parlour Press that the present editors have been able to trace is represented in the photographs and plans that follow. That it is one of the original presses made by Holtzapffel is shown by an inscription stamped into the frisket frame: COWPERS PARLOUR PRINTING PRESS / HOLTZAPFFEL & Co LONDON / MANUFACTURERS & SOLE AGENTS. The only other marking on the press is the number 940 punched into the end of the bed, between the hinges. Its date of manufacture is uncertain, but seems likely to have been between 1850 and 1880.

The press itself is made of mahogany, the bed and platen being $1\frac{7}{8}$ inches thick by $11\frac{1}{4}$ by $7\frac{15}{16}$. The top of the platen stands 7 inches from the table. The overall size of the press when closed, including handle, hinges, and frisket thumb-piece, is height about 11 inches, width $8\frac{3}{4}$, and length about $14\frac{1}{4}$. The hinges are very strong, being of cast iron with right-angle pieces half an inch thick, set well into the bed and platen. The eye against which the bridle pulls is almost half an inch in diameter and springs from a plate of similar thickness screwed into the underside of the bed. The plate on top of the platen, on which the toe of the lever works, is a little over $\frac{3}{8}$ inch thick. The lever itself is $\frac{7}{16}$ inch thick, $9\frac{7}{8}$ inches in length to the end of the tang, or $11\frac{7}{8}$ inches to the end of the rosewood handle; it works within a bridle made of iron a little over $\frac{7}{16}$ inch in diameter. The screws employed on the hinges are size 14, $1\frac{1}{4}$ inches long, countersunk, with very slightly rounded heads. The inner dimensions of the bed are approximately $9\frac{1}{2}$ by 8, and allowing for wedges would accommodate a chase with inner dimensions of approximately 8 by 7. The frisket is apparently short, but in fact allows for the wedges and thickness of the chase and will permit a full use of the bed: its inner dimensions are 8 by $6\frac{1}{4}$. Although it is of thicker metal at the hinge end ($\frac{5}{32}$ as against a little under $\frac{1}{8}$ inch), it is not in fact tapered as would be usual with a larger press. The tympan, which is screwed onto the face of the platen, is larger than the frisket, measuring externally 8 by $10\frac{1}{16}$ inches, internally $8\frac{7}{8}$ by $6\frac{3}{8}$; the existing tympan frame, which however looks as if it may not be the original, has been cut out of $\frac{1}{16}$ inch iron sheet.

There are rebates cut in the base pieces and a strip of wood screwed along the far edge as a stop, to accommodate a small drawer ($7\frac{7}{8}$ wide by $7\frac{5}{8}$ by perhaps $1\frac{1}{2}$ inches) which is unfortunately no longer present. The press is therefore presumably one of the more expensive ones described in the price list, page 78 of the manual, as 'japanned and finished in the best manner, and fitted with a drawer'.

The accompanying plans of the press now in the possession of David Chambers were drawn full size and are reproduced to a scale of 1 : 4. The photographs were taken by Richard Brown.

1:4 SCALE DRAWINGS OF A PARLOUR PRINTING PRESS

a bed; *b* platen; *c* lever; *d* bridle; *e* plate on which toe of lever bears; *f* tympan; *g* frisket; *h* hinge; *i* handle of lever

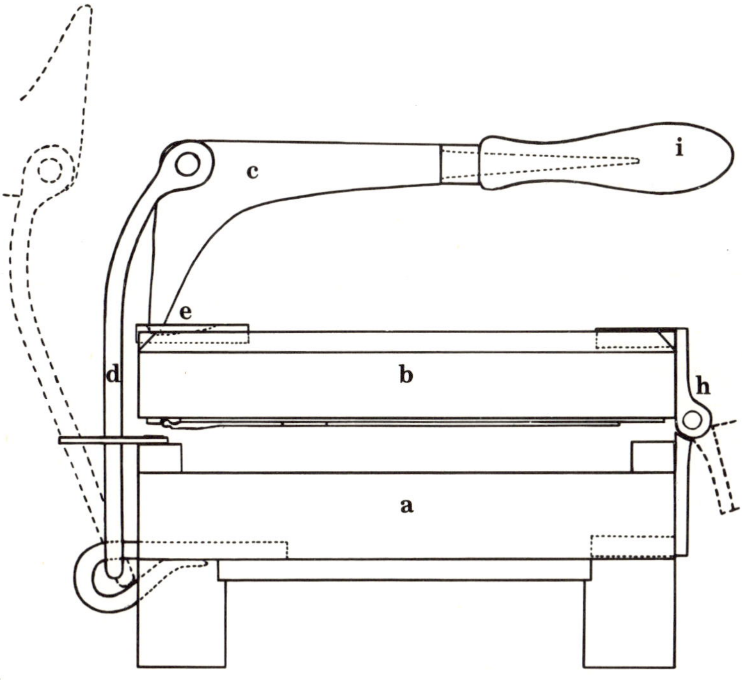

Fig. 10. Side view of press, closed.

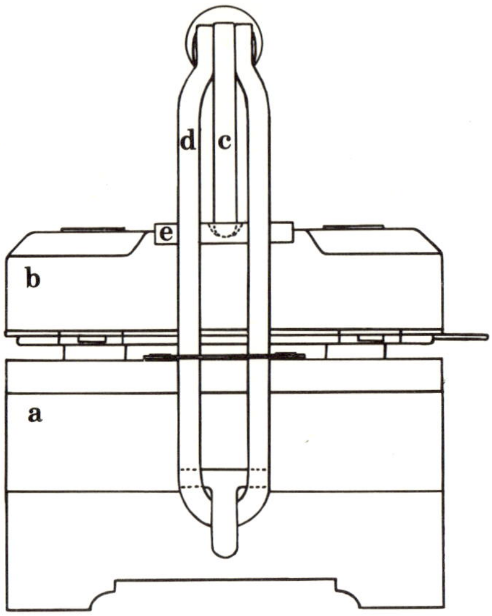

Fig. 11. Front view of press, closed.

Fig. 12. Plan of bed, open.

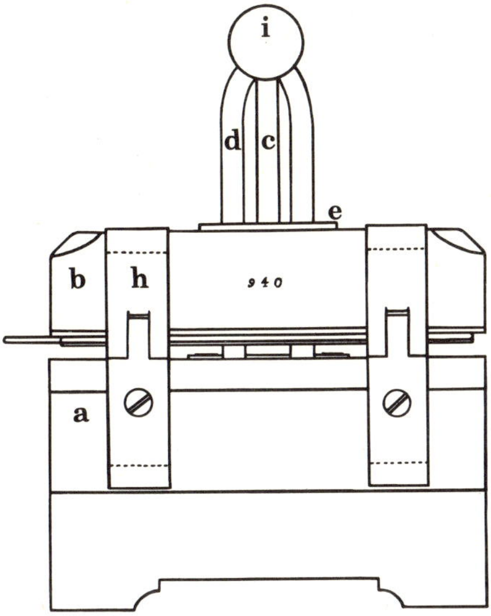

Fig. 13. Rear view of press, closed.

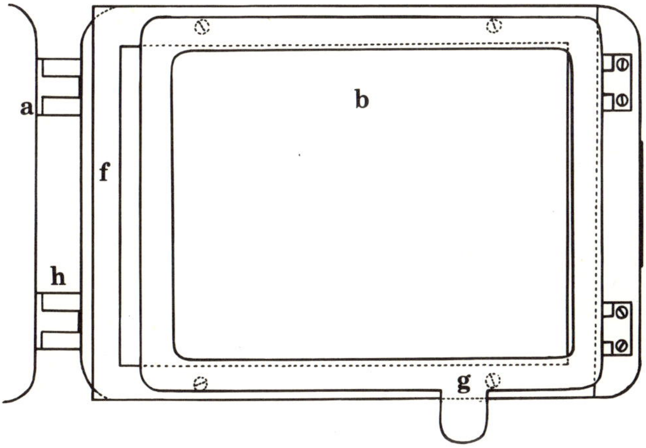

Fig. 14. Plan of platen, open.

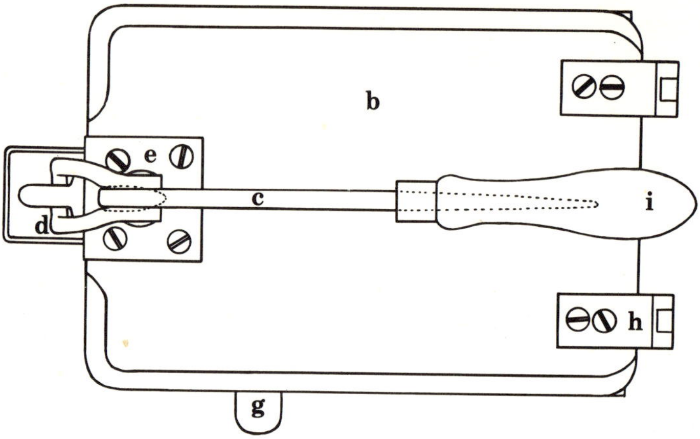

Fig. 15. Plan of press, closed.

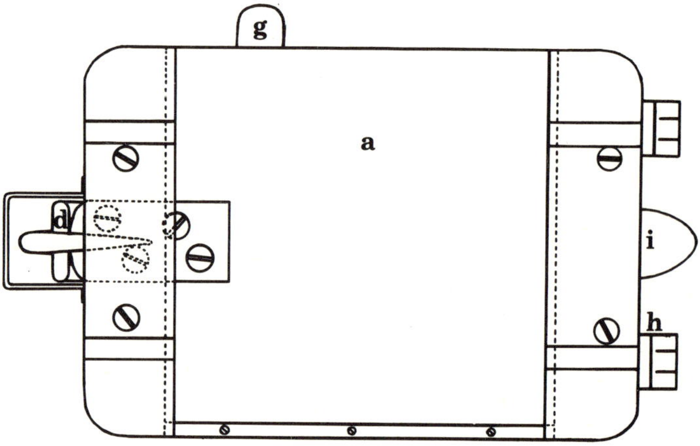

Fig. 16. Plan of press, closed, from below.

Plate 1. One of Cowper's Parlour Printing Presses, fully opened.

Plate 2. The press, closed, from the front.

Plate 3. The press, closed, from the rear.

Plate 4. The press, closed, seen from below.

THE FACSIMILE

Notes on the text will be found at the end. Their existence is indicated by the following sign in the margin ▷

PRINTING APPARATUS

FOR THE

Use of Amateurs.

CONTAINING

FULL AND PRACTICAL INSTRUCTIONS

FOR THE USE OF

COWPER'S PARLOUR PRINTING PRESS.

ALSO THE DESCRIPTION OF

LARGER PRESSES ON THE SAME PRINCIPLE, AND VARIOUS OTHER APPARATUS FOR THE AMATEUR TYPOGRAPHER.

THE WHOLE MANUFACTURED AND SOLD ONLY BY

HOLTZAPFFEL & Co.,

ENGINE, LATHE, AND TOOL MANUFACTURERS, LONDON.

THE PAMPHLET CONTAINS LIKEWISE, NUMEROUS SPECIMENS OF PLAIN AND ORNAMENTAL TYPES, BRASS RULES, CHECKS, BORDERS, ORNAMENTS, CORNERS, ARMS, &C. &C.

THIRD EDITION, GREATLY ENLARGED.

PUBLISHED BY HOLTZAPFFEL & Co.,

64, CHARING CROSS, AND 127, LONG ACRE, LONDON;

AND TO BE HAD OF ALL BOOKSELLERS.

1846.

LONDON:
BRADBURY AND EVANS, PRINTERS, WHITEFRIARS.

TABLE OF CONTENTS.

Section Three.

Section Four.

PRINTING APPARATUS

FOR THE

USE OF AMATEURS.

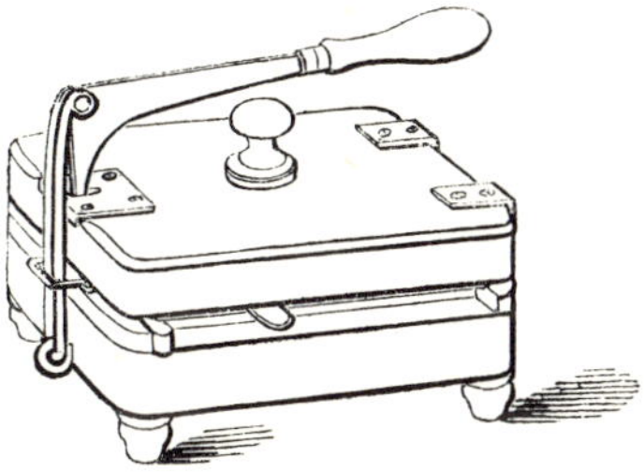

INTRODUCTION.

THE Parlour printing press was invented by Mr. Cowper for the amusement and education of youth, by enabling them to print any little subject they had previously written, provided the printing did not exceed in size the dimensions of an ordinary duodecimo page, which measures about 5 inches by 3 inches.

The press itself may be considered to have owed its origin to a small and old instrument, known amongst printers as the *Bellows Press*, that has been occasionally used in former days for printing cards, little papers and bills; and in which instrument, two small but thick rectangular boards, terminating in handles, are hinged together, somewhat as in the familiar household implement, from which the bellows press derives its name.

A very superior modification of this press was constructed in metal, by the celebrated Timothy Bramah, for printing the dates and consecutive numbers on the Bank of England Notes, after the general impressions had been printed from a copper plate. This bank note numbering machine is still employed; it shifts the figure at every impression, and answers very well because the types to be imprinted on the note, or rather on two notes united end to end, lie in one long line, and for which one line, the press is abundantly powerful.

But when it becomes the question of printing an extended surface of four or five inches square in this manner, although the pressure is sufficiently near to the hinges of the old-fashioned bellows press, the opposite end of the impression fails, because an appropriate pressure cannot be there given by the simple lever hitherto employed, as the pressure continually decreases from the hinges to the handle.

Taking up the contrivance at this point, Mr. Cowper conceived the idea of the removal of the former handles of the bellows press altogether, and producing the pressure at the open end, by a modification of the well-known mechanical combination, the *toggle* or *knee-joint*, as represented in the wood cuts, pages 5 & 10; and therefore in Mr. Cowper's Parlour Printing Press the pressure at the one end of the press is due to the hinges, and at the other end to the toggle-joint, and with a little care the two actions may be so nicely balanced, that the pressure on the surface may be made strictly uniform.

Mr. Cowper also designed the very ingenious box with drawers, that serves to contain the appropriate number of types for the parlour printing press; these types consist of numerous varieties, and which, unless kept in *strictly orderly arrangement*, combined with great *facility of access* would be nearly useless; these necessities have been completely met, in a box of a very compact and convenient form, which is also arranged to contain in a like miniature form, the roller, inking table, and other adjuncts required in every application of the typographic art.

The appropriateness of these several arrangements to their common purpose, may be considered established by the complete satisfaction they have conveyed to all those who have tried the apparatus, success having been invariably attained in their use.*

Cowper's parlour printing press, as before observed, was first intended only for the amusement and education of youth, for which purpose it is not to be easily surpassed, as it supplies an interesting recreation requiring both exactness and method; but

* Mr. Cowper's talents have been successfully employed in designing both extremes of printing machines, from the colossal "Times" newspaper machine, which prints 4200 sheets in the hour, or 70 papers in one minute, measuring 48 inches by 36 inches, to the present miniature and fire-side apparatus.

experience has proved that the parlour press is equally useful for numerous purposes of a more extended character. Thus, for example, companies, institutions and individuals, have found it convenient for circular letters, invoices, and papers subservient to the despatch and methodical arrangement of business; naturalists and travellers, for short memoirs of scientific researches, or labels for specimens; the principals of scholastic establishments, for the multiplication of various papers employed in conducting the classes and studies of their pupils; and amongst the pupils themselves for small essays, original papers and school newspapers. The parlour printing press has also been found useful to clergymen and numerous others, interested in the spread of religious knowledge and the management of charitable and savings funds; also to many of the friends of education, for disseminating original and other papers; to wood-engravers for examining the progress of their blocks; to practical printers for proofs of title-pages, stereotype plates, or cards; and indeed almost every different pursuit will suggest some new application of this useful little apparatus.

The parlour printing press was first intended to be used alone for little pages not exceeding at the most about four or five inches square, but the manufacturers have tried the same construction with complete success for presses having several times that area, or sufficiently large for printing the half-sheet of demy paper, which measures 18 by 11 inches; beyond this the Stanhope, Columbian, and other printing presses, although more cumbrous and weighty in structure, are more convenient in use.

It has been occasionally doubted by those who have not examined the mathematical principles of the toggle-joint, knee-joint, or the perpetual lever, by which several names it is alike known, whether it could give the required amount of power. The larger application of the principle just referred to, may fully silence these doubts, which may also be set at rest by a reference to the grand and simple test of all mechanical combinations, intended to increase the power actually applied, namely, that "what is lost in motion is gained in power," in other words, by comparing on the one hand, the space moved through by the hand in producing the effect, with, on the other hand, the space moved through by the open end of the press, where the effect or gain is to be estimated.

On examination it will be seen that whilst the hand moves the lever of the press through the entire range of its compression, or about five inches, that the front of the press only descends about one quarter of an inch, thus making during the entire period, an aggregate gain of 20 times the power employed. But during the last *one inch* of the movement of the lever, the descent of the press is so very small a quantity, as to be quite inappreciable to the eye; assuming however, that the press descends the one hundredth of an inch, whilst the handle is moved the last inch, the aggregate power for this last space moved through by the hand, would be one hundred fold the primary force. But further, as the power of the lever continually increases from the beginning to the end of its action, at which latter period the two parts of the toggle-joint come *nearly into one right line*, and at which the effect of the lever is considered mathematically to cease; the power at the very last point of the action, is enormous, perhaps a thousand-fold, and has given to the construction the very appropriate name of the *perpetual* or *infinite* lever, from the absolute movement of the press or other machine, dwindling down to a quantity almost ideal or infinite, and the comparative power, rising in a corresponding manner almost to infinity.

In simple fact, although the effective pressure is so considerable, a very small primary force is required in its production, as a child ten years old can efficiently use the parlour printing press. In the larger presses the power required is also very little, and in these latter the first exhibition of feebleness is not in the toggle-joint, but in the springing or bending outwards of the small but thick boards of which the presses are constructed; those of the *folio foolscap* press, measuring only 10 inches wide and 15 inches long, but 2½ inches *thick;* those of the *half-sheet demy* press, measuring only 12 inches wide, and 20 inches long, but 3 inches *thick.*

▷ In these large presses it is very desirable the bed on which the types stand should have a cast-iron plate, planed accurately flat, a construction more permanently accurate and trustworthy, than the wooden bed; and in these large presses it is also convenient that the top board should be counterpoised, which greatly facilitates the freedom and facility with which the larger presses may be used.

The general principles of the press having been noticed, the reader is referred to the four sections of which this pamphlet consists, the heads of which are as follow:—

The first section, contains a description of the portable and economical printing apparatus more usually supplied by Holtzapffel and Co. to beginners, and which suffices for producing a small page of printed matter not exceeding six inches square. An apparatus of the same size and capability, but of a more handsome form is also spoken of. The first section of the pamphlet is concluded by full instructions for using either of these portable or parlour printing presses.

The second section, describes a larger apparatus on the same general construction as the parlour press, and adapted to printing pages of any size not exceeding that of half a sheet of foolscap paper. This description is followed by instructions for the use of such parts of the larger apparatus as differ from those of the parlour press. Instructions are subsequently given for printing in two colours, and in bronzes; a few words are added on printing music by letter-press, and also on making composition inking rollers and dabbers.

The third section, refers to various trays and cases for containing either small or large quantities of type in orderly arrangement, which is followed by a description of the hand chase for printing impressions not exceeding three inches long by one inch wide, applicable to the printing of endorsements, titles, labels, notes, &c., upon papers, books, or drawings, to which purposes ordinary printing presses are in general inapplicable. The monotype printing press is then figured and described, and Professor Willis's scheme for producing architectural plans, &c., by a novel application of monotype printing, is also briefly noticed.

The fourth and last section, contains specimens of various types, comprising a selection of Roman, Italic, Black Letter, Script, and various ornamental types, and also ornamental dashes, borders, corners, brass rules, &c., but any other kind of types beyond those inserted will be supplied to order, namely types for the Greek, Hebrew, Saxon, and numerous Oriental Languages, and types for mathematical signs and characters, the whole of which may be used in Cowper's Parlour Press, or the Folio Foolscap Press.

Section I.

COWPER'S PARLOUR PRINTING PRESS AND APPARATUS.

Fig. 2.

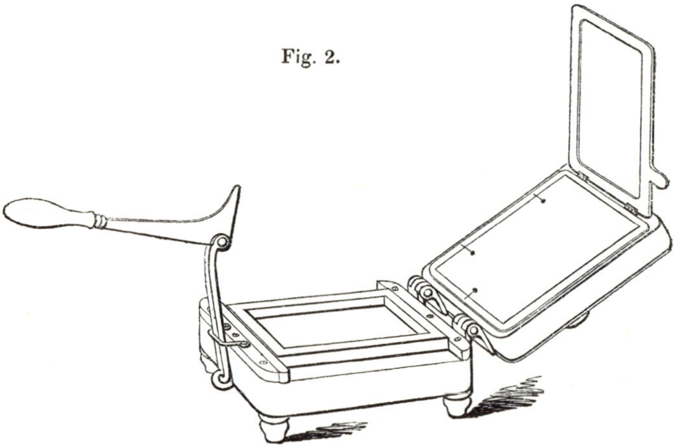

Description of the Parlour Press.

Cowper's parlour printing press, shown in the above wood-cut, is made of mahogany and stands in the small space of 11 inches by 8, and 10 inches high. The press is quite portable, it is neat in appearance, and sufficiently large for a sheet of post or letter paper, as it is capable of printing a surface 7 inches long and 6 inches wide, independently of the margin. In printing a page of this size the press works so easily that a child 10 years old may use it on the parlour table without difficulty.

The press, which is extremely simple in its general construction, consists of two stout blocks of mahogany; the lower piece called the *bed* rests upon feet, and to the back of the bed is hinged the upper piece called the *platten*, which closes upon the bed nearly as a lid shuts upon a box, but leaving a parallel opening between the bed of the press and the platten, just equal to the height of the types when '*set up*', or arranged for printing from, in the rectangular iron frame called a *chase*.

The pressure is applied to the front of the press by a lever, which is jointed to the upper extremity of a long bridle, the lower end of which is hinged to the bed. When the platten is shut down, the lever is immediately turned over and acts by a species of knee-joint or toggle-joint, in which the leverage is constantly increasing during the descent of the lever, until at the conclusion it becomes almost infinite. In this manner a considerable pressure is obtained by the exercise of a very slight force; and when the lever has completed its office, the space between the platten and bed being quite parallel, the pressure is uniformly distributed over the entire surface of the types; notwithstanding that from the angular action of the press it may appear to be more powerful near the hinges, which particulars have been fully explained.

If however, the hard surface of the platten were permitted to press upon the types, the faces of the letters would be injured, and the impression would not be uniform; the platten is therefore covered with a piece of thick woollen cloth called the *blanket*, and to render the surface smooth the blanket is itself covered with a piece of parchment called the *tympan*. A light square iron frame, or '*frisket*' is hinged to the face of the platten, and serves to keep the paper in its position whilst it is being printed.

To print a page of types that has been set up in the galley chase, and inked, the sheet of paper is placed on the tympan, the frisket is turned down to hold the paper, the platten is then closed upon the types, and the lever is brought over until the press assumes the position shown on page 5; the impression is then completed, and it only remains to open the press and take out the printed paper. The entire process of printing with the parlour press is both easy and expeditious, as will be hereafter more fully and practically explained.

Description of the Small Deal Type Case, Type, &c.

The type case that accompanies the plain parlour press is made of deal, and painted; its external measurement is 14 inches long, 10½ wide and 6 in. deep, the case has four drawers suitably partitioned to receive the type and furniture. A fount or series of about 2500 Pica Roman types is contained in cells or '*boxes*' ◁ in three of the drawers of the type case, every cell or box in the drawers is made of a size proportionate to the number of types it has to receive, and the front of every cell is made sloping, in order that the types may be readily drawn out with the forefinger when they are required for composing the words of any subject. The arrangement of the drawers and boxes is shown in fig. 4, page 19.

The supply of type comprises large capitals, figures, double letters, small letters, (called lower-case letters,) points or stops, spaces, quadrats, and space line leads, amounting in all to 96 varieties, or '*sorts.*' This selection of type contains the most general kinds, and includes all such as are absolutely required in the first commencement. A great variety of additional characters are made to every series or fount of type, so that in many cases the complete fount consists of upwards of 300 varieties; but the great majority of these are but seldom used, and then only for peculiar purposes which vary in almost every instance according to the nature of the subject printed. It has therefore been found advisable to limit the first supply, to the types required in every case, leaving those few kinds that may be necessary for particular purposes, to be added as occasion for them may arise. This arrangement avoids needless expense, and prevents the inconvenience of having a large number of varieties, or '*sorts*' that might never be required.

The quantity of type contained in the case, as usually supplied, will be found adequate to printing a page about 6 inches by 3, if the type be set up moderately close. But if it should be required to print a page with the lines of type quite close together, and of the full size that the press is capable of receiving, namely about 6 inches by 7, it will be found necessary to have a duplicate set of type. The type case is made sufficiently large to contain the double set of types should it be required, the remainder of which can therefore be added at any subsequent period.

The number of types of each letter, furnished in a complete fount, varies according to the recurrence of the letters in ordinary composition; thus in a fount of types to which there are twelve thousand of the small *e*, the proportion usually allowed to the *z*, is only two hundred, or sixty of the *e*, to one of the *z*, and so on of all the other letters. This method of proportioning the sorts or kinds, is adopted in order that one sort may not be exhausted much before the others. In the type supplied with the parlour press, the numbers of the letters are proportioned in the same method, but somewhat modified so as to render the supply as generally useful as possible.

Fig. 3 represents an ordinary type, the technical names of the several parts of which are as follows:—

L. The *letter*, standing upon

S. The square piece of metal called the *shank*.

W. The *width*, which varies much in different letters, as in the I and W for instance.

B. The measure on this side is called the *body*; it is always the same for every letter in the same fount, (or set of types,) to cause every letter to stand in an exact line.

F F. The *foot* of the type.

C. The *channel* or groove; it is at this point that the metal is poured into the type mould, and when the waste piece of metal called the "break" is broken off, the burr that is left is planed away and forms this groove.

Fig. 3.

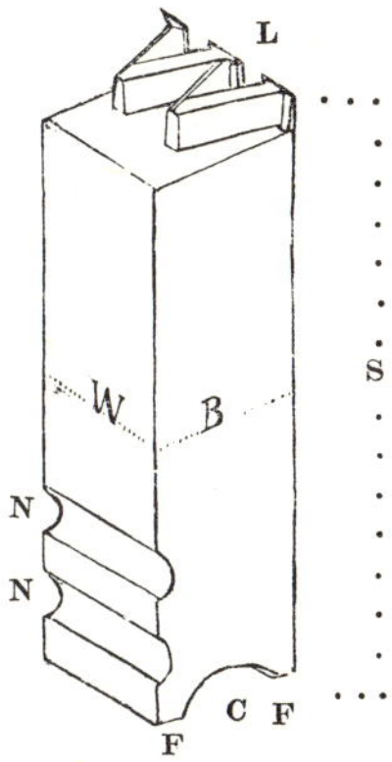

N N. The *nicks;* these are to show the compositor which is the proper way for the type to stand, and prevents the necessity of looking at every individual letter, for he can feel with his thumb whether he has it the right way upwards; there are sometimes two, three, or four nicks cast in the type, the number and distances of the nicks being intended to distinguish one fount from another.

Spaces and quadrats, are blank types or pieces of metal, used for separating the words of type from each other, and also for leaving such blank spaces as may be required at the ends of lines. The spaces and quadrats only differ from each other in their thickness, and are known as, hair spaces, thin spaces, thick spaces, *en* quadrats, *em* quadrats, 2 *em*, 3 *em*, and 4 *em* quadrats; the hair spaces are made very thin, and the 4 em quadrats measure about $\frac{1}{2}$ an inch in the same direction.

The spaces and quadrats are all made of the same body as the fount of type to which they belong, so that they may stand exactly in a line with the type. In fact if the type shown in fig. 3 were shortened about a quarter of an inch from the upper end, by cutting off the letter, it would represent an *em* quadrat, which size is exactly square, or has the width and body alike. The *em* quadrat forms the standard for all the other sizes; thus the *en* quadrat is half the thickness of the *em* quadrat; the thick space one third, the thin space one fifth, and the hair space about one tenth the thickness of the *em* quadrat; the two, three, and four *em* quadrats, as their names imply, are respectively two, three, and four times the width of the *em* quadrat. Thus the spaces and quadrats, from their graduated differences in size, admit of being so combined as to fill up with great exactness any blank spaces that may be required amidst words, or at the ends of paragraphs; and further particulars of their respective uses will be found under the head Composing.

Space line leads, are thin pieces of type metal used to separate the lines of type, when they are required to stand wider apart than the space given by the body of the type. An instance of their use may be seen in almost every newspaper, where an article may perhaps, for half its length, have the lines of type sufficiently wide apart to be read quite easily, and the remainder of the article may have the lines very close or somewhat crowded together. The same types are in this case used throughout, but where the lines are wide, space line leads are introduced, and where the lines are close, the leads are omitted, thus leaving the lines of type immediately in contact with each other, or as it is called '*solid;*' this is done in order that the whole article may be compressed into the allotted space by the removal of part of the space line leads, to contract the column in the required degree.

In a similar manner, it is usual in almost every case to separate the lines of type more or less according to circumstances, and in order that any requisite separation may be thus made between each line of type, the leads are made of all thicknesses, namely from 2 leads to the thickness of one pica type, to 12 or 14 in the same thickness; they are designated as 2 to pica, or 12 to pica as the case may be. The space line leads supplied with the type case, are of the thickness of '4 to pica;' that is, four leads are equal to the body of pica type. This size will be found useful for most purposes, and one or more may be put between each line of type, to spread the lines in the required degree.

Among type founders and printers, the Pica type is always made the unit, or foundation, of all their measures; not only are the space line leads denominated as 2, 3, 4 or 12 to pica as they are the half, third, fourth or twelfth the thickness of the body of pica type, but large types for posting bills are called 5, 12 or 30 lines pica, when their height is five, twelve, or thirty times the height of one line of pica type; and the width and length of a page of printed matter is never estimated in inches, but as 30, 40, or 100 *ems* wide or long. In fact the printer has no professional acquaintance with feet and inches in his craft, but employs as his unit of measure the *em* pica, of which about seventy-one are equal to one foot of ordinary measure.

To return from this digression, it is to be observed as of importance that the length of the leads should be exactly equal to that of the line of type, or the width of the page or column; because if the leads were a trifle too short, it would be found very difficult to make the last types of each line keep their proper positions, and if the leads were too long it would be necessary to fill out the length with spaces or quadrats placed at the end. Three inches is the length of the majority of the space line leads in the type case, that being found the most useful size, some six inch leads are also included in the case, and two of the three inch leads may be placed end to end to make six inches, but should an intermediate length be required it would be advisable to have space line leads of the appropriate size.

In addition to the types and space line leads above mentioned, the type case contains the following articles:—A supply of wooden furniture,—brass rules,—a box of printing ink,—an inking roller,—and a zinc tray to serve as an inking tablet; these also require some few words in the way of description.

Wooden furniture, may be considered to include all the pieces of wood used for securing the type in the *chase*, or the rectangular frame in which the types are set up ready to be printed from, and these several pieces of wood are known as reglet, furniture, side sticks, foot sticks, and quoins.

Reglets, are parallel pieces of wood of the same thickness and height as the quadrats, but are generally cut to the same length as the space line leads, and are used in the same manner, for separating such lines of type as are wished to stand wide apart; the reglets are also sometimes used to make up the length of the page.

Furniture, are pieces of wood thicker than the reglet, and are used for the same purposes, but chiefly for extending the mass of types to the full length of a page, or so nearly to fill up the space between the type and the chase, that the side sticks and quoins may be brought into use. Furniture is also frequently used to separate two pages of type from each other, so as to form the margin between the pages, when printed and folded.

Side sticks, *foot sticks*, and *quoins*, are wedges of various lengths used for securing the type in the chase in the manner shown in fig. 6, page 23. The side stick is placed against the side of the page of type, and should be a trifle longer than the page, the foot stick should be equal to the width of the page but not longer, or it will be very liable to catch against the side stick. Quoins are wedges about one inch long, and of various thicknesses, that some of them may be found to suit the opening left between the side sticks and the inside of the chase; and when the quoins are driven in firmly, they are said to '*lock up the form*,' which latter is the technical name for the page of type when wedged fast in the chase, ready to be printed from, as seen on page 23.

Brass Rules are for printing single, double, or waved lines, as shown on page 74.

Composition inking roller, *box of ink*, and *inking tablet*. These are fitted in the upper drawer of the type case, and will be found well adapted to their united purpose, namely, that of inking the type preparatory to printing the impression. The inking roller is a cylinder made of a composition of glue and treacle, as explained at page 50—53: the roller is three inches long and one inch in diameter, and is fitted into a small iron frame that is sprung open to allow of its introduction. The roller will be found to ink the types more uniformly and rapidly than the '*dabbers*,' or '*balls*' sometimes used for the same purpose. The box of ink and inking tray are too simple to require description.

Description of the Set of Extras.

The set of extras, comprises the tools that are required in using the parlour printing press, they are modified from the corresponding tools used by printers, to suit the convenience of amateurs: their names and uses are as follows:—

▷ *Transfer composing stick*, shown in fig. 5, page 19, is used in arranging the type, an operation called composing, the types are first '*set up*' or placed in their proper sequence in the composing stick, and are afterwards transferred to the chase in the manner explained under the head Composing, pages 18—22. The transfer composing stick is intended to receive only one line of type at a time, and has a false bottom, the withdrawal of which deposits the line of type in its place, without risk of accident, and obviates the chief difficulty encountered in the first attempt at composing.

Bodkin. This is a slender pointed awl, in a hard-wood handle with sheath. In correcting the errors made in composing, the bodkin is used to pick out such of the types as are misplaced.

Forceps. These are of the ordinary kind, and are used for removing such types as cannot be so easily drawn out with the bodkin.

Mallet. This tool is made of hard wood and used both with the shooting-stick and planer.

Planer. This is a piece of hard mahogany about 4½ by 3 inches, made flat on one side and round on the other, and used to level the surface of the page of type, which is done by placing the flat side of the planer firmly upon the type, and striking the round side with the mallet, this operation is called planing, and is performed when the types are only partially fixed or wedged up in the chase.

Shooting-stick. This is a wedge-shaped piece of hard wood, used with the mallet to drive the quoins up tight; this process is called '*locking up.*' The shooting-stick is placed with its smaller end against the quoins, whilst the larger end receives the blows of the mallet.

Brush and Turpentine. These are used for cleaning the type, as explained on page 32.

Two quires of demy printing paper, cut to the sizes most useful for the parlour press, one quire being of the octavo size, and one quire of the full size the press will admit.

Pair of damping slates. Before printing on the paper it is necessary it should be wetted, after which it is placed between the slates, to keep it flat, and distribute the moisture equally. The particulars of this trifling but *important* process, will be hereafter given on page 25.

Large galley chase. This is an iron frame, measuring internally 7½ inches square, and fitted with an iron bottom. When the frame has no bottom it is called simply a *chase*, and it is invariably used in this condition by printers; but the addition of the bottom, which converts it into a *galley* chase, prevents the types from falling through when they may have been imperfectly locked up, an accident that might frequently happen to beginners with the ordinary chase. The large galley chase is of the full size the parlour press will admit, and is used for printing all pages exceeding 6 inches by 3; below that size, the small galley chase will be found the more convenient.

The last few pages having conveyed the complete description of the parlour printing press and its apparatus, some instructions for their use will next be offered.

INSTRUCTIONS FOR THE USE OF THE PARLOUR PRINTING APPARATUS.

In the following instructions, for the management of the parlour printing press and the apparatus connected with it, the object attempted is, to explain the most simple and easy mode of arranging the type, and working off the finished impressions, in such a manner as may be readily understood by persons having no previous knowledge of the subject; and this without regard to the mode generally followed by practical printers, who require a quantity of means and apparatus too extensive for the purposes of the amateur.

The order of description will be the same as the order of proceeding necessary to produce a number of perfect impressions, and will commence with the first step towards the arrangement of the type, called '*composing*,' and end with the last process required after the impressions have been produced, namely the replacement of the type in the cases ready for future operations, and which process is called '*distributing*.'

Composing.

The small type cases each contain four drawers numbered consecutively, the upper drawer No. 1 contains the furniture, and the lower three drawers contain the type. For composing, it will be found most convenient to arrange the drawers in the order shown in fig. 4, as then the capitals, or upper-case letters, will occupy the upper left hand space, the figures will be immediately beneath them, and the double letters close to the two drawers containing the small or lower-case letters, the cells or boxes of which letters will be found to run alphabetically, the spaces and qnadrats will be to the left hand, the stops and space line leads to the right.

The galley chase intended to hold the type when composed ready for printing, should not be placed level, but inclined in two directions, so that the right hand corner the furthest from the operator may be the lowest; this may be done either by resting its front and left hand edges upon two books, or pieces of wood, or the galley chase may be placed obliquely upon an inclined board, or on a writing desk turned round, so as to make the fur-

Fig. 4.

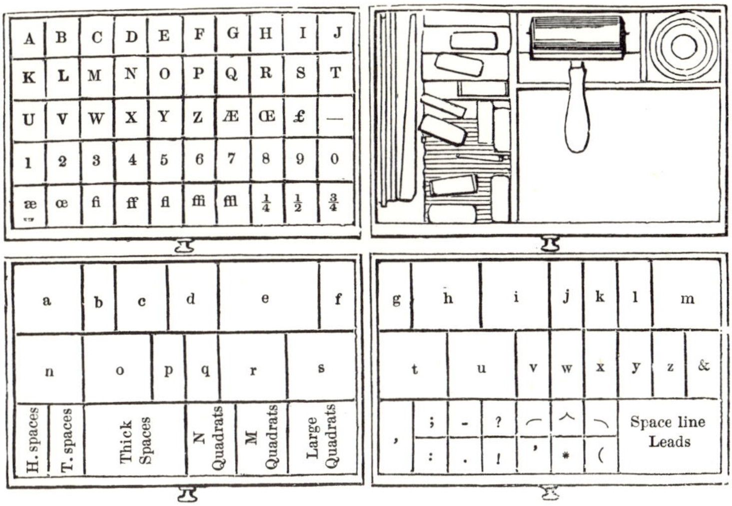

ther edge the lower. If the galley chase is placed in this position, all the types will lean towards the right hand corner, and support each other, but if the chase is placed level, considerable annoyance may arise from the type falling. The inclined galley, described on page 39, will be found a very convenient support for the chase, giving it the required inclination; it is sometimes supplied along with the smaller or parlour presses, as well as with the foolscap press which it usually accompanies.

The transfer composing stick fig. 5, is to be held in the left hand, with the right hand corner the lowest, as represented, this inclination is adopted as in the galley chase, to prevent the types falling. A space line lead should be placed in the composing stick, and all will be then ready for composing, as follows:—

Fig. 5.

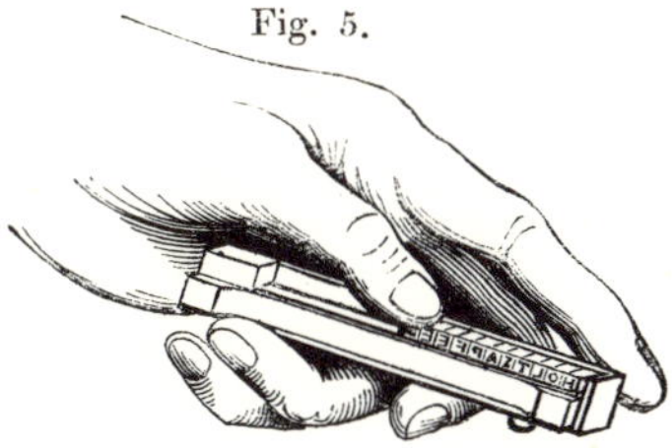

With the forefinger and thumb of the right hand, take out of the case the first letter of the matter to be set up, place it in the right hand corner of the transfer composing stick, and press it into its place with the thumb of the left hand as shown in fig. 5, observing always to keep *downwards* the side of the type having the nicks in it, and therefore the side of the type seen upwards should always be *smooth*, this must be particularly attended to, as otherwise the letter when printed from, will be inverted.

Each succeeding type required to form the word, must be added in the same manner, and in order to economise the time occupied in composing, the left thumb should be raised ready to receive the type, immediately that the right hand brings the type to the composing stick, the type is then to be pushed into its place with the end of the left thumb, during the time the right hand is employed in obtaining the succeeding type; and so on with each letter until the word is completed, a *thick space* should then be placed at the end of the word, that space being proper to form the division between the words, which are all to be set up in the same manner. When the line of type is nearly completed it will in almost every case be found, that the line is either a little too long, or a little too short, to fit the composing stick without dividing the last word in the midst of a syllable, and which must of course be avoided; the adjustment of the line of type to the length of the composing stick is called '*justifying*,' and is accomplished in the following manner.

If the line of type is too long, the thick space between each of the words may be taken out and a thinner substituted. If the line is too short, a thicker space may be used between each word; and in this manner the length of the line may be justified until it exactly fits the composing stick.

It will however frequently happen that taking out all the thick spaces in the line, and substituting thin ones, will make the line too short, and it may be only necessary to alter two or three spaces to gain the desired room; in this case the space should be taken from between those words which terminate in *round* letters, such as *c*, or *e*, because these may be brought closer together without appearing crowded, which is not the case with *upright* letters such as *d* or *b*. Indeed two consecutive words that respectively terminate and commence with such upright letters, may with advantage always have a space between them somewhat thicker than that between words with round letters; and therefore when the line requires to be very slightly lengthened, preference is given to putting thicker spaces between the upright letters, so as to maintain an equal appearance of space or '*white*' between every word. For the same reason it is good practice in the first setting up of the type, always to place an *en* quadrat, (which exceeds the measure of a thick space,) after any word ending with the letter *f*, the upper part of which letter overhangs the body of the type, and therefore requires a thicker space after it, to leave the same amount of white as between other words.

Care should be taken in justifying, that each line fits the composing stick equally tight, should this not be attended to, those lines that fit tight in the composing stick, will from being longer than the others be held tight when the page is locked up, but the shorter lines will be loose, and from having a slight rolling or lateral motion will print imperfectly. Scarcely any amount of after-labour will remedy this evil, and therefore care should be taken in the first instance, to make every line fit the composing stick equally but moderately tight. Much time will also be saved if the words are carefully examined as every line is set up, to detect any errors of spelling or composition. With a little practice this requires scarcely more than a glance, and any error may then be easily rectified, whereas if neglected in the first composition, much additional trouble will be afterwards necessary in the correction.

The line of type having been set up and justified in the transfer composing stick, the next operation is to transfer the line of type into the galley chase; to effect which, a piece of brass rule should be placed on the front of the line of type, it will then be as it were in a square box, and if the brass rule be held tight in its place, the types may be carried, in the composing stick to the chase, without any risk of their falling.

To deposit the type in the chase, the brass rule and transfer composing stick are grasped together, between the forefinger and thumb of each hand; the whole is then placed in the galley chase close to the right hand side, in the position the type is intended to occupy; the two wooden knobs are then taken hold of with the thumb and forefinger of the left hand, which press the type close against the front of the chase, but without exerting any pressure downwards. Whilst the left hand is thus pressing the type forward, the thumb and forefinger of the right hand, grasp the brass knobs and withdraw the entire piece of metal forming the false bottom; the type then drops into its proper position; the remaining part of the transfer composing stick, and lastly the brass rule are removed without difficulty, leaving the space line lead standing below the level of type. After one or two trials the operation will be found very easy, and to occupy far less time than the reading of the above description.

The remaining lines to constitute the page, are added in precisely the same manner, and attention to the following quotation will be found very much to abridge the time occupied in setting up the types.

"In composing, the left hand which contains the stick should always follow the right which takes up the letters. If the left hand remain stationary, much time is lost in bringing each letter to it, and traversing a greater space than is necessary; the eye should always be fixed on the nick of the letter before the finger is ready to take it up, this will effectually prevent any false motion, as it may be lifted and conveyed to the stick in its proper position. A sentence of the copy should, if possible, be taken at one time, and while putting in the point and space that conclude that sentence, the eye is at full liberty to revert again to the copy for a fresh one."*

Locking up the Form.

When the page of type has been composed, the next process is to secure it firmly in the chase by means of the wooden furniture, so that no one type may be in the least degree loose, nor any one project in the slightest manner above its fellows, as that would destroy the uniformity or exact level of the types, which is quite essential to good printing. When the type is securely fixed in its chase, the whole collectively is called a '*form*,' and fig. 6 shows the appearance of the form when finished. Arranging the wooden furniture around the type is called '*dressing the form*,' fixing the type in the chases is called '*locking up*,' and making the surface of the type level is called '*planing*,' the three processes are carried on as follows.

To dress the form, one of the thin slips of wood called *reglet*, is placed between the right hand side of the page of types and the side of the chase, this reglet should be at least equal in

---

* Hansard.

length to the page of type and is pressed closely against it; should the page be a short one it will be necessary to place

Fig. 6.

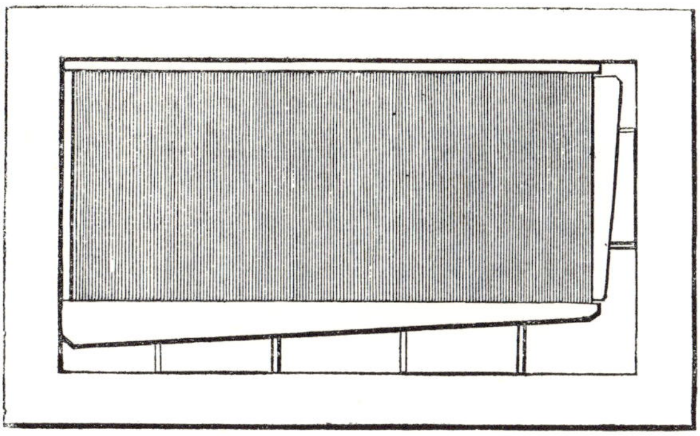

some of the reglets, or some of the wider pieces called furniture, at the ends of the page to extend the length to about 6½ inches. This will leave room for the foot stick which is placed at the bottom of the page, and also for the one or two small wedges called quoins; the side stick is placed at the left hand side of the page, and two quoins are fitted to it, the form is then said to be dressed.

To lock up and plane the form, the quoins are first thrust up firmly with the thumb, and then tightened by placing the end of the shooting-stick against each quoin, and slightly driving it forward by blows of the mallet. The flat face of the planer is next placed firmly upon the face of the type, and a few steady blows of the mallet are given upon the round side of the planer, which is slightly shifted between each blow so as to traverse it over the entire surface of the type in succession.

When the surface of the type has been thus rendered level, the shooting-stick is again had recourse to, the quoins are driven in tight, and the planer is also again used, to ensure a level surface. The form will be then complete as shown in fig. 6, and ready to be taken to the press for printing the first impression, called '*taking a proof*,' a step preparatory to correcting the errors that may have been made in the composing. For particulars of the mode of taking a proof, the reader is referred to page 27.

Correcting the Form.

Notwithstanding all the pains that may have been taken in the composing, it will be found in almost every instance that some errors have been made, and these have now to be corrected. For this purpose the proof sheet is carefully examined, rather spelling the words than reading them. The errors most likely to escape observation are the substitution of the letter *c*, for the *e*, and of the *n*, for the *u*, or the reverse, and also the inversion of the letter *s*, these letters therefore require particular notice, and all the required corrections should be carefully marked with a pen upon the proof sheet. The quoins around the type are next loosened, and the first wrong letter is drawn out either with the bodkin or forceps as most convenient; the proper letter being substituted immediately after the wrong one has been withdrawn. When the bodkin is used for this purpose it is gently thrust into the shoulder or side of the type, by which a sufficiently secure hold is obtained to lift the letter.

To avoid confusion, the corrections should be commenced from the head of the page, working regularly downwards, and should a whole word or more have to be removed, it will be found best to raise the entire line about half way up, by placing the forefinger of each hand against the ends of the line of type and lifting it altogether, the word may then be drawn out without injuring the type with the bodkin.

When the corrections have been made, the form is again ▷ locked up and planed, a second proof is then taken and compared with the first to see if the corrections have been properly made, the second proof should also be carefully read, and if any errors still remain after the '*revise*' or second proof has been examined, they are to be corrected in the manner above directed until the work is apparently faultless. When all the corrections have been made, and the type is locked up and planed, the business of composition is finished and the form is said to be '*ready for press.*'

PREPARATIONS FOR PRINTING.

The first step towards printing the impressions is to prepare the paper, which should be of the kind known by stationers as *printing* paper; this is manufactured with considerably less size on its surface than writing paper, and it therefore receives the ink much better. The paper supplied in the set of extras, is printing paper of a good quality, and is cut to the sizes most suitable to the two chases, but of course, paper of any size within the limits of the press may be used.

It is in all cases requisite to wet the paper before use, as the printing ink will then be found to penetrate the pores of the damp and softened paper, which will therefore receive a good impression, and when the paper has been dried in the air, the ink will in general be found to be permanently fixed. On the contrary, printing ink does not so thoroughly attach itself to dry paper, and therefore a comparatively imperfect impression is produced in the first instance, and the ink then requires so much longer time to dry, that the impression is liable to be smeared or to set off at every touch for some days after having been printed.

Damping the Paper.

To damp the paper a pan or large basin three parts filled with clean cold water may be used; about six or eight sheets of paper, are then taken loosely between the thumb and fore finger of each hand, in such a manner that the paper bends downwards, in a segment of about one third of a circle. The right hand is then dipped obliquely into the water, and the paper gradually drawn under the surface, so that the whole extent of the paper may be successively and equally immersed. As the hands are removed from the water, the paper is allowed to assume the vertical position for a moment that the surplus water may run off. The six sheets of paper thus wetted are then laid on one of the damping slates and about an equal number of dry sheets are laid upon them. Another half-dozen sheets are then wetted and these again covered with about six dry sheets, and so on alternately until the whole '*heap*' of paper is '*wetted down*,' observing that the half-dozen sheets at the top of the heap are always to be wetted. The second damping slate is then laid on the paper, and any convenient weight of from 20 to 50 lbs. is placed upon it, and in a few hours the paper will be ready to be printed upon.

It is usual to damp the paper the evening before it is wanted, that it may stand the whole night to allow the moisture to become equally distributed, but if carefully and equally damped, the paper may be used after the lapse of about a couple of hours.

Should the paper be very stout it is drawn through the water slowly, and should it be very thin a greater number of sheets are taken and they are drawn through the water more rapidly. When properly damped and ready for use, the paper will lie flat and feel damp to the touch, but no moisture will be visible to the eye. Should not the paper be damp enough, a few sheets may be again wetted and distributed equally throughout the heap, and should the paper be too damp, a few dry sheets distributed in like manner throughout and allowed to remain an hour or two will rectify the accident.

The exact degree in which paper should be damped depends jointly upon the quality of the ink and on that of the paper itself, for if a *soft spongy* paper with much moisture in it is used with *strong* ink, small portions of the paper will stick to the type in printing and be torn from the surface, leaving the impressions indistinct and the type *foul*, owing to the particles of paper left upon it; and therefore it will be found that the thicker the ink and the softer or thinner the paper, the less moisture the latter should contain, but a little experience will soon give the requisite information on this point.

Distributing the Ink.

In this process a small quantity of printing ink should be taken out of the box with a knife or spatula, and spread in a narrow line across the end of the inking tray. The roller is then to be placed lightly upon the ink, *rolled* backwards to the other end of the inking tray, and then *lifted* from the surface so as to be *carried*, and not *rolled* back, to the ink.

This operation if repeated five or six times, will distribute a sufficient quantity of ink over the surfaces of the roller and the inking tray.

To distribute the ink still more uniformly, the roller may be next rolled backwards and forwards several times, both straight along and obliquely across, until the ink forms a thin uniform layer over the inking tray. If in the first instance the roller be *drawn* backwards and forwards, without being *lifted* up and carried back, the same part of the roller would constantly come in contact with the body of ink, and the result would be that instead of being evenly distributed, the ink would be accumulated in black patches, this may be entirely avoided by lifting the roller up and carrying it back between each stroke as explained.

Inking the Types.

For this purpose the inked roller is to be rolled *very* lightly across the page of types, not from head to foot, but from side to side, so that the roller may travel in the direction of the lines of type. In first putting the roller upon the type, it is usual to place it in the *middle* of the width of the page, as then the exact height at which the roller should be held, is at once felt, whereas if the rolling be commenced from the *edge*, the outer types will in general receive too much ink.

For the first impression the roller should be rolled four or five times backwards and forwards over the types, for the subsequent impressions twice will be found sufficient.

If the roller be shorter than the page it should first be placed at the head, next moved to the middle, and then to the foot of the page, so as to ink the type equally all over, this should be particularly attended to, as otherwise the impressions will not be of uniform blackness or colour.

Taking a Rough Proof.

But little preparation is necessary for taking a rough proof, or printing the first impression by which the errors of composition are detected. The form or assemblage of type is placed on the press in such a position, that the page of type is central with the press, and therefore the chase is somewhat on one side; the form is then secured in its position with the long wedge as shown in fig. 2 page 10.

The sheet of paper is placed centrally on the tympan, and the frisket is closed upon it, and after the type is inked the platten is turned down, the lever is next brought over, and keeping the bridle away from the front of the press, the toe of the lever is placed in the recess in the square iron plate. The handle of the lever is then depressed until the bridle comes in contact with the front of the press, just as it is represented on page 5; the entire force of the press will now have been exerted, the lever may be immediately returned and the press opened for the removal of the proof. The whole process of closing and opening the press will be found very simple and easy after one or two trials, and not to require more than four or five seconds of time.

Working off the Impressions.

For working off the entire number of impressions, a little more preparation is requisite than for taking a proof, and the first step is to *cover the frisket.* For this purpose, a sheet of stout paper about three inches wider and longer than the frisket is damped and pasted around its four edges, it is then laid upon the parchment tympan, the frisket is turned down upon it and the pasted edges of the paper are folded around the frisket and secured to the central part of the paper just within the margin.

When this paper or the *frisket-sheet* is dry, an impression of the types is taken on the frisket-sheet, and the impression is then cut out, to make an aperture just large enough to permit the type to print upon the sheet of paper, when it is laid upon the tympan and beneath the frisket-sheet, which latter, as it only exposes the precise spot of the paper to be printed upon, prevents the remainder of the same from being soiled in the operation, besides serving to keep the sheet in its position, should it be smaller than the frisket.

▷ The *tympan-sheet* is next fixed: for this purpose a sheet of damp paper of the same size as those to be printed, is slightly pasted around the edges, and fixed on the parchment tympan in such a position that the printing may come, as nearly as it can be judged of, in the middle of the paper. The press is then closed upon the type as in printing, except that, as the object is merely to obtain a barely visible impression, the ink left upon the type when the last proof was pulled will be sufficient for the purpose.

If the impression should fortunately happen to be quite central upon the fixed or tympan-sheet, it only remains to thrust two common pins in contact with the sides of the tympan-sheet, and in a slanting direction into the blanket beneath, allowing the half of the bodies of the pins to stand out; and a third pin is similarly fixed half-way in the blanket, and in contact with the foot of the page. These three pins will serve as guides for the position of all the future sheets, as if every one of these is slid under the projecting ends of the pins so far as these will allow, each sheet will be placed in the same relative position to the type, and when printed upon will show exactly the same equality of margin throughout the entire quantity.

It is however far more likely the sheet first pasted on will not have been so judiciously placed as to receive a strictly central impression, and in this case a clean paper of the same size can be used to measure the amount of the displacement, and show how distant from the exact edge of the fixed or tympan-paper the pins should be inserted; and the pins if needful must be shifted by trial until at length the impression shows an exactly equal margin at the head and foot of the page, and also at the two sides. The true position of the pins having been thus found, and which presents little or no difficulty, the whole quantity of paper may be worked off without apprehension of error, provided every sheet is put close up to the pins.

At the commencement of printing it is always desirable to draw the thumb-nail diagonally four or five times over the top sheet of the heap of paper, to cause the few upper sheets to project slightly; and before drawing off each sheet as it is wanted in the working, the thumb-nail is similarly drawn once over the top sheet; this serves continually to loosen the upper sheets, which gradually creep forward in detached order, so as to enable the operator to take up a single sheet with considerable facility.

In forming a judgment of the uniformity of the impression it is usual to examine the *back of the sheet*, because if the indentations of the type upon the paper be equal, there is nothing more to be desired in the condition of the press, and the faults, if any, arise from the inking having been irregularly performed. But if the indentations are uneven, then those portions which are faint should be carefully cut out from the impression and pasted upon the corresponding parts of the tympan sheet; these pieces are called '*overlays*,' and give additional pressure where added.

When wood blocks are printed from it is usual to put the required thicknesses of paper or card *under* the blocks to cause them to agree in height with the types, and these pieces are called '*underlays*.' In either of these modes, or more conveniently by a combination of the two, the impression from any form, or from any wood block, however irregular, may be made sufficiently equal throughout for the purpose of printing by the employment of *overlays* and *underlays*; but it saves much trouble when the wood blocks exactly agree in height with the types.

Should the impression, although equal, be too faint, notwithstanding that a sufficiency of ink has been used, it may be remedied by placing one or two thicknesses of card either under the entire surface of the chase, or under the blanket, as may be found the most convenient. When from having been long in use the press constantly requires this addition of card, it will be found best to loosen the hinges from the platten of the press, and also the square piece upon which the lever acts, and to place the required thickness under all three alike; the hinges and lever plate are of course to be screwed firmly in their places immediately after.

Practical printers sometimes prefer to place the overlays *under* the parchment tympan ; this may be readily done in the parlour press by withdrawing the four screws which fix the tympan to the platten, adding the requisite thickness of paper, and afterwards refixing the tympan. In the same manner the blanket may be removed when necessary, either to soften it by rubbing it between the hands, or at other times when it may be desirable to exchange the blanket for one either thicker and softer, or thinner and finer. Small type closely arranged requires a soft thick blanket, and large type widely separated requires a thin ▷ blanket, or sometimes even a piece of kerseymere cloth is alone employed, and at other times combinations of the two qualities of cloth or blanket are resorted to; this must, however, be left to the judgment and experience of the operator.

When the tympan becomes loose from the stretching of the parchment, the latter should be renewed; this may be easily effected by cutting a piece of parchment about two inches longer and wider than the tympan frame, pasting it around the edges with thick paste, then laying the tympan frame on the parchment and folding the edges around the sides and ends of the frame, tucking or thrusting in the edges thus turned over; lastly, the central part of the parchment is wetted to allow it to stretch, and the edges are then drawn tight. This mode allows the paste to dry first, thoroughly secures the parchment, and leaves the covering of the aperture or panel in one single thickness, which on drying will be found perfectly tense if the covering have been carefully managed.

Sometimes one part of a page may have too much pressure upon it, as for instance, when the one-half consists of small type close together, and the other half of larger types widely separated; it may then happen that the pressure just suffices for the end of the page at which the small type is situated, but is too great for the end of the page at which the lines of type are distant.

In this case, practical printers remedy the evil by the introduction of '*bearers*,' which serve to reduce the pressure upon the types, or rather to bear off a part of the pressure; this is usually done by pasting one side of a piece of reglet of appropriate thickness, and laying it with the pasted side upwards, upon a part of the furniture immediately adjoining that part of the page which is too strongly pressed upon; the first impression that is afterwards taken fixes the piece of reglet to the frisket, as the latter is strongly pressed upon the pasted side of the reglet, and which latter '*bears off*' the platten from that part of the page.

The thickness of the piece of reglet should be such that when placed upon the furniture it may rise a little higher than the face of the type. Sometimes pieces of cork are used, which possess the advantage of elasticity, and more commonly a roll of paper is pasted on the frisket in the same manner as the reglet. ◁

Should the amateur require to print two impressions on opposite sides of the same piece of paper, as the two pages on the one leaf of a book, it may be accurately accomplished, by the use of '*register points*,' described on page 41, and which points may be also applied to the parlour press.

Drying and Pressing the Paper.

When the whole of the impressions have been printed off, they are spread on a table or hung on poles or lines to dry, after which the paper is pressed, to make it flat and remove the indentations caused by the printing. For this purpose printers' bookbinders and others use a powerful screw press, or a hydrostatic press, and from 50 to 500 sheets are placed between two flat boards and the whole are firmly squeezed in the press by the united exertions of two or more men. The amateur will however in general be able to succeed very well with an ordinary letter-copying press, by placing the paper between milled boards. A common table linen press will also partially answer the same purpose. In both these cases it will however be found best, not to press more than 100 sheets at a time, and if half a dozen glazed milled boards are distributed amongst these, the pressing will be the better done. When the work is pressed in a hurry and before fully dry it is usual to put a glazed board between every sheet, to prevent the *setting off*, and this is always done in hot-pressing, in which process heated plates of iron are inserted at small intervals, amidst the pile of paper that is being hot-pressed.

In the absence of any of these means, the paper may be made moderately level by placing it between two flat boards, or even between the damping slates, and loading them as heavily as convenient for a few hours. Some ladies have accomplished the pressing with a heated laundry iron, taking three or four sheets together, with blank papers interposed. In which way soever the pressing is done, it is necessary to be careful that the paper is nearly dry, as otherwise there is danger of the ink upon one sheet being transferred to the adjoining one, this which is called '*setting off*,' would materially injure the appearance of the work if not altogether spoil it.

Cleaning the Type.

To return to the consideration of the type from which the impressions have been worked, as soon as the printing is completed the form should be removed from the press and the ink thoroughly cleaned from the type, this may be done either with turpentine, or with a solution of pearl-ash called '*lye*.'

If turpentine is used, it is only necessary either to pour a few drops on the type, or to dip the brush into a shallow vessel containing it, such as a plate, and then to apply the brush with moderate force to the entire surface of the type; this will immediately remove the ink, and the turpentine may be soaked up by dabbing and pressing a cloth upon the face of the type.

When however the type is frequently cleaned with turpentine, it is apt to become sticky from some of the turpentine being left upon the type, and it is therefore occasionally necessary to clean the type with pearl-ash, and which is perhaps the better mode at all times, notwithstanding that it is somewhat more tedious.

To make the lye, half a pound of American pearl-ash should be added to three pints of soft water, this may be kept in a covered jar to be always ready for use. To clean the form, the brush should be repeatedly dipped into the solution and scrubbed freely over the type, after which the type should be thoroughly washed with clean water, until the whole of the pearl-ash is removed, as should any remain, the alkali would corrode the type. When perfectly well washed, the quoins should be loosened the type again rinsed and allowed to become nearly dry, after which it may be distributed for future use. Should the pages have been left firmly locked up after the washing, and allowed to remain until perfectly dry, the types would be liable to stick together and cause trouble in their separation.

Cleaning the Roller.

The roller perhaps more than any other part of the printing apparatus requires to be kept scrupulously clean, for should any extraneous matter lodge upon it, or any dry ink remain from previous use, the next time the roller is employed for inking the type the extraneous matter will be transferred to the type, and by clogging up the letters produce black patches in the impression. It is therefore necessary to wash the roller every time after use, and which if regularly done will keep it in good condition as long as it will last; with care this will be about three months, as in the course of that time the roller, even if scarcely used, generally becomes too hard for further service, and a new roller is then requisite.

To clean the roller it is only necessary to wash it with *cold* water and a piece of linen rag. If the roller is covered with ink, little or no impression will at first be made upon the roller, as the ink being ground in oil is not acted on by the water, but in about two minutes the water will dissolve a thin film of the composition of which the roller is made, and the whole of the ink will then wash off, leaving an entirely new surface on the roller, which should be then rinsed clean, and hung up by the frame to dry for about two hours, by which time it will be ready for use. Care should be taken to avoid touching the surface of the roller after it has been rinsed, until it is quite dry, as when wet the surface of the roller is soft and liable to injury.

If the roller, although clean, should be found too hard, it may be softened by soaking it in cold water for about half an hour, and afterwards holding it within a foot of a moderate fire for a few minutes, taking care to turn the roller round to expose all sides equally to the warmth, and prevent any one part becoming over-heated; it must be remembered, that if soaked too long in the water, or held too near the fire, the roller will partially dissolve and become useless; for the same reason the roller should not be washed in warm water, except indeed in very cold weather.

The inking tray should also be cleaned after use with the turpentine or pearl-ash in the same manner as the types. It is indeed quite essential to good printing, that perfect cleanliness should be observed in the types, roller, and inking tablet; and any pains that may be bestowed with this object, will yield an ample return in the beauty of the printing, and the ease with which it will be accomplished, whereas if all three, or even one of them be foul, it is impossible to obtain a clear impression.

Distributing the Type.

Distributing the type into the cases ready for future use is exactly the reverse of composing, and is pursued in the following manner; after the form has been unlocked, the side sticks and furniture removed, and the type placed with the head of the page the highest.

The first word of the page is taken up, with the thumb and second finger of the right hand applied at the two ends of the word, and the fore finger placed over the middle of the word, to assist in separating it from the rest of the type; in this manner it is placed between the insides of the thumb and fore finger of the left hand, which are held straight. The types are then taken separately with the thumb and fore finger of the right hand and gently thrown into their respective boxes, observing however that each letter need not be looked at individually, but that the entire word is noticed at the time it is lifted from the page, and in the process of distributing, the word is *unspelt*, beginning with the first letter of the word, by these means the necessity of closely examining every letter is avoided, and the operation is rendered very much more rapid and certain.

The recollection of the sequence of the words prevents the slightest difficulty in reading the type, even to a first beginner, and if one or at most two words be taken at a time, the lifting up and distributing will be found very easy, and much more rapid than it may at first appear to be.

In distributing the type, it is recommended always to take hold of each type by the *letter* and throw it gently into the case *foot foremost;* by following this rule the risk of bruising the face of the letter is nearly avoided, and the types are placed in the most convenient position for picking them up when they are next composed. In distributing the spaces, a little care is required to avoid mixing those of different thicknesses; these must be individually examined, which will however present little difficulty.

Much time will be saved in future operations by careful distribution, as if it is done hurriedly, wrong letters are put into the respective boxes, which causes the future composition to be imperfect, and vastly increases the labour of correction.

Having now completed the detailed account of every step to be followed in using the parlour printing press, it only remains to be added to this section, that the foregoing descriptions apply alike to the plain press, and to the best parlour printing press and apparatus, which contain precisely the same number of parts, and only differ in respect to their general appearance and exterior finish.—The following pages refer to additional printing apparatus, for amateurs, subsequently introduced by Holtzapffel & Co.

Section II.—FOLIO FOOLSCAP PRINTING PRESS AND APPARATUS,

ON THE PRINCIPLE OF COWPER'S PARLOUR PRESS, BUT MADE OF INCREASED SIZE.

These apparatus are larger than the parlour presses, &c., and admit of more extended employment; but some modifications have been made in their construction both to suit this increased size, and to assimilate the arrangements to those usually adopted by practical printers. The parts of the apparatus that differ from those supplied with the parlour press will be now described, leaving unnoticed such of the means and methods as are similar in both instances, and which have therefore been already adverted to.

Description of the Folio Foolscap Press.

This measures externally 21 by 10 inches, and 12 in height, the bed of the press measures 15 by 10 inches, and the limit of the surface of type that can be printed in it, independently of margin, is 12 by 7½ inches. This is only ½ an inch shorter and narrower than the half-sheet of foolscap; the press is therefore abundantly large for that sized paper, which may be made to contain one, two, or four pages of type according to circumstances. The platten and frisket of the press are each provided with stops to prevent their being turned back too far, the press is also provided with a pair of '*register points*,' which are used as guides when the paper has to be printed on both sides, to ensure the second page being placed exactly on the back of the first or '*in register*,' and the two chases supplied with the press are not fitted with bottoms, but are left open as employed by practical printers. In taking impressions the folio foolscap press is worked in exactly the manner already described, as being applicable to the parlour press.

Description of the Large Deal Type Case.

This measures externally 24 by 18 inches, and 12 inches high; it is furnished with handles, a lock and key, and contains six drawers, four of which are partitioned after the printers' method for holding type, and two for receiving tools, furniture, &c.; the types in the four lower drawers comprise upwards of 9000 pieces and include—

Great Primer Roman, namely, capitals, figures, points, spaces, and quadrats. See specimen No. 9, page 65.

Pica Roman, large and small capitals, lower case (or small letters,) with accented vowels for printing the foreign languages, figures, points, spaces, and quadrats complete. No. 13.

Bourgeois Roman, namely, capitals, figures, points, spaces, and quadrats. No. 17.

Bourgeois Antique, namely, capitals, figures, points, spaces, and quadrats. No. 23.

The arrangement of the type in the drawers is exactly the same as that adopted by printers generally. The Pica Roman being the principal fount, requires two drawers, called respectively upper and lower case, from the position they occupy on the composing frame, as shown on page 38. The upper case is equally divided into 98 boxes, and receives the large and small capitals, figures and accented letters. The lower case is divided into 53 boxes of four different sizes and receives the small letters, points and spaces. The size of each box is proportioned to the number of types it has to contain; thus the box for the letter *e* is the largest, and that for the *z* one of the smallest; these compartments exactly agree in size and arrangement with those of the ordinary printing-office.

The arrangement of the boxes is not alphabetical, but that which brings the letters of most frequent recurrence close together, and thereby lessens the distance the hand has to travel in composing ordinary language; much time is saved by this arrangement of the boxes, and although at first sight it may appear somewhat complicated, the positions of the various letters in the lower case will be remembered with even greater facility than the places of those in the upper case, although the types are there arranged almost alphabetically.

The small letters from being placed in the lower case, are themselves always called 'lower case;' (thus it is usual to say lower case *e*, *l*, *m*, &c.,) whereas the type in the upper case is generally described as large capitals, (E, L, M,)—or small capitals, (E, L, M).

The Bourgeois Roman being only a small fount without small capitals, is contained in one drawer, the capitals on one side and the lower case letters on the other.

The Bourgeois Antique capitals and figures, and the Great Primer Roman capitals are both contained in one drawer, the lower case letters not being added in the box, to either of these kinds.

The fifth drawer contains a proportionate supply of space line leads, furniture, reglet, side sticks, foot sticks, and quoins.

The upper drawer, which is 3 inches deep, contains the necessary tools for printing, made of a size suited to the foolscap press; they consist of a mallet, shooting stick, planer, printer's composing stick 9½ inches long, bodkin, brush for cleaning the type, a pair of thick damping slates, and a slate to serve as an inking tablet. These tools are similar to those described on page 17, except the composing stick, which is described pages 39—41.

Composition Inking Roller in Frame and Case.

This roller is a cylinder of composition 6 inches long and 2 inches diameter, cast upon an inner core of wood perforated through its axis and bushed with brass. The frame is made of iron, and the roller is secured in the frame by an iron spindle that passes through the axis of the cylinder and is fixed by a screw and nut. The handle of the frame has, in place of a ferrule, a disk about 2½ inches diameter, this prevents the hands from being inked by the roller, and also keeps the handle raised from the table so that it may be readily grasped.

The case for the roller is of deal, 7 inches square and 3 deep, and is so fitted, that when the roller is placed in the case and the lid secured, the frame of the roller is fixed, and that without the surface of the roller being liable to injury from accidental pressure, which would indent and spoil it.

It is desirable that the roller for inking the types should be a little longer than the page of type, as it saves time, and the inking is somewhat easier; and for this reason a 12 inch roller is occasionally used with the folio foolscap press.

Composing Frame.

This is a frame for supporting the type cases in a convenient position for reaching the types. The printers use for this purpose a frame about 4 feet 6 inches high, of about the same length, and adapted to holding four cases (as the trays containing the types are technically called;) the two lower cases are placed at an angle of 20 degrees, and the two upper at an angle of 55; this has the effect of bringing the whole of the types easily within reach of the compositor, who usually stands at his work.

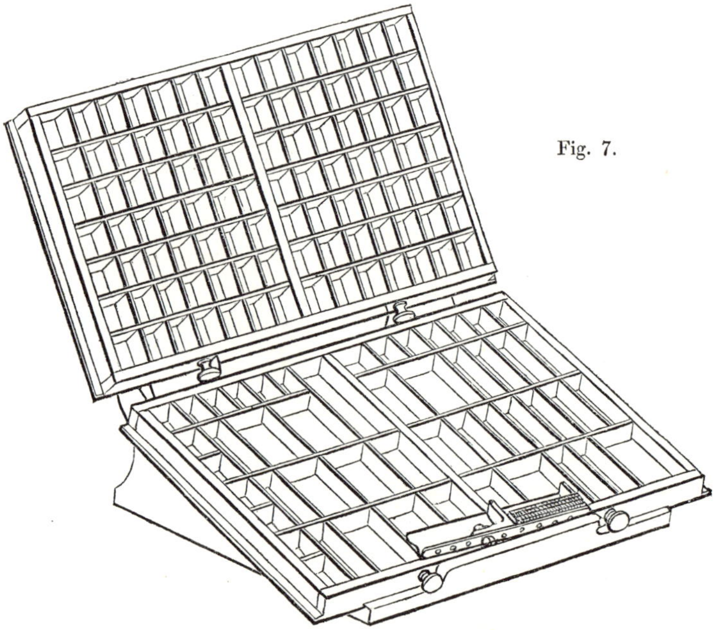

Fig. 7.

Such an arrangement would however be too cumbersome for most amateurs, and therefore a composing frame is supplied that measures only 20 by 18 inches and 14 inches high. This will support 2 drawers of the large type case at the customary angles, and when placed upon an ordinary table, the height will be found such as to admit of the amateur compositor being seated. Fig. 7 shows the position of the two drawers when placed upon the composing frame ready for operations.

Inclined Galley.

The ordinary galley used by printers is simply a thin board with ledges on one side and end, and not inclined: on this galley the compositor places the types he has composed, and when he has set up a page, he ties a string round it, and lifts the whole page of type in his fingers with little or no risk of accident.

As however this might be found too difficult for the amateur the inclined galley is made to stand at an angle of 10 degrees from the horizontal, and with the right hand corner the lowest, it is also provided with a thin taper moveable board, and upon this the types are placed. When the page of type is set up, a string should be tied four or fives times round it, the page of type should be then drawn into the middle of the thin moveable board of the inclined galley, upon which it may be readily carried about. To remove the type from the board it is only necessary to place the latter flat upon the spot where it is desired the type should be deposited, and then, whilst holding the type in its place with the left hand, to draw the taper board from beneath the type with the right hand, this is very easy and entails no risk of accident.

Instructions for the Use of the Folio Foolscap Press and Apparatus.

The main differences in the use of the parlour and the folio foolscap apparatus are—in the composing sticks—in tying up and imposing the two or more pages of type for the larger press—and in the use of the register points,—these several matters will therefore be next described in the order in which they are above mentioned.

The ordinary printer's composing stick is employed with the larger type case, because the quantity of type to be set up is greater, and in general those amateurs who use the larger press, will have gained some experience in the management of types from using the smaller press, so as to enable them to use the printer's composing stick without much difficulty; whereas had this instrument been employed at first, it is probable some amateurs might have met with failures and disappointments and have thereby become discouraged.

The printer's composing stick fig. 8, page 40, is a trough of sheet iron or brass, with an L formed sliding piece *a*, which can be fixed at any distance from the head *b*, according to the width of page required; the width having been determined, and space line leads provided of the requisite length, a piece of brass rule of the same height as the type, is cut to the exact length of the intended lines and with a small projecting knib at the one end, this is called a *setting rule*, and is placed in the composing stick

Fig. 8.

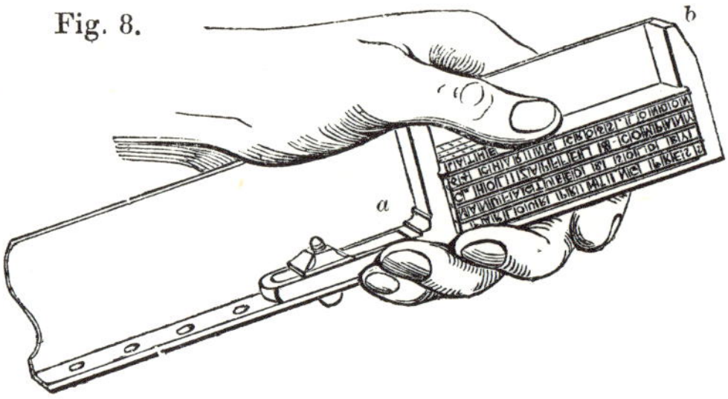

which is held in the left hand, with the thumb reaching over the slide and resting on the setting rule as represented. It will be observed that in using this composing stick the inclination is exactly the reverse of that requisite for the transfer composing stick described on page 19, and in like manner the direction in which the types are placed is also reversed; that is, in the printer's composing stick the nick of the letter is always placed upwards or in sight, and the commencement is made from the left hand side instead of the right.

One line of type is first composed and justified, a space line lead is placed over it, and the setting rule is lastly lifted out by its knib and placed upon the line of type just composed; a second line is then set up, and so on until about seven or eight lines have been put in the composing stick, this quantity which is as much as the stick will conveniently hold is generally called a '*stickfull*.'

The next operation is to remove the type from the composing stick to the galley, for this purpose the printer generally places the setting rule over the last line of type, and grasps the type firmly with both hands, the two forefingers being placed over the setting rule, the two thumbs over the space line lead in front of the types, and the sides of the middle fingers being pressed tightly against the ends of the lines; in this manner the type is usually lifted out bodily, and carried in the same manner to the galley.

The amateur will however be more certain of success, if before grasping the type, he places the composing stick upon the table, sideways but nearly vertical, and resting against the front of his person, the two forefingers may then be slid under the space line lead in front of the type, the thumbs placed over the setting rule, and as the type is drawn out of the composing stick the middle fingers are to be pressed upon the ends of the lines. By this method the types will always lie on their sides which are supported by the space line leads, and the danger of the types slipping through the fingers will be avoided.

In the manner above described the type may be easily carried to the galley, in which it is placed as near its position as possible and with the forefingers touching the bottom of the galley. The whole is then turned over to bring the types upright, grasping the mass firmly at the moment and turning it over steadily, but so rapidly that the type has scarcely time to fall. In practice this will be found very easy, and if the first attempts are made with only two or three lines of type success is almost certain; the number of lines may be gradually increased as the operator becomes more dexterous.

Each stickfull of type is added in the same manner to the matter previously composed, until enough has been set up to form a page, which should then be tied up and removed to the press. To tie up the page, a piece of fine twine is very tightly wrapped four or five times around the page of type, and secured at the right hand corner either by tying it firmly, or by folding the end of the string three or four times around the coil, and drawing the end tight. The page of type may then be removed as explained at page 39, and placed either on the press or upon a stout sheet of paper, which latter enables the page of type to be afterwards moved with facility.

If any danger is apprehended in carrying the page of type, a little water may be applied upon it with a sponge, the capillary attraction will cause the water to become disseminated amidst the types, and make them so far to adhere, that with ordinary care an accident can scarcely occur in their removal.

Register Points.

When both sides of a sheet of paper have to be printed, it is of importance that the one impression should be placed exactly on the back of the other, so that if the sheet be held up between the eye and the light no difference in their positions may be perceptible. To attain this exactness, a pair of register points is used, each point consists of a steel wire rivetted into a shank about 3 inches long, made of thin steel and having a groove, for the reception of the screw by which it is secured to the platten. The groove admits of any position being given to the point within certain limits, and by which abundance of range is obtained.

In all cases when the paper is *backed*, or has two impressions, the form, frisket, &c. for the first side are made ready, and pins are placed for giving the position of the impression on the first side as formerly directed at pages 28 to 29, the register points are then passed under the blanket and thrust through the tympan just within the margin of the sheet of paper to be printed, and each at about an equal distance from the center of the page; the register points are then secured firmly in their places by the same screws that fix the tympan; some prefer to fix the register points *above* the blanket and tympan, and it is desirable that the points be fixed near the top of the sheet, and opposite each other or square across the page. The act of printing the first side, then causes two holes to be likewise pricked by the points, exactly in the same relative position to the impression on every sheet of paper, notwithstanding that the paper itself may vary in size, or be imperfectly placed in the first working.

In making ready for printing the second side, the position of the chase should be carefully adjusted, so that when the paper is turned over and so placed that the register pins exactly enter the holes formerly made in the paper, the second side may appear truly on the back of the first when the sheet is held between the eye and the light. The necessary exactness having been obtained by trial on one or more sheets, the chase is then secured, and if the indentations of all the remaining sheets are in like manner placed upon the register points, the same accuracy will be thus obtained throughout. Should it be more convenient, the position of the register points instead of that of the chase, may be adjusted for obtaining the true register of the second side.

Imposing, or Arranging several Pages to be Worked Together.

The instructions offered at pages 22-3 on locking up the form were intended principally to refer to those cases in which but one impression or page is printed on each sheet of paper, or at most when a second impression was added at the back, which is called *backing*, or working the *reiteration*. But if four or eight pages are to be printed on one sheet of paper at two impressions, one on each side, it becomes necessary first to set up the whole four or eight pages, and then to arrange them in two forms, the one for the outside of the sheet or '*the outer form*,' the other for the inside or '*the inner form*,' so that when both sides of the sheet of paper are printed, and it is folded, the pages may follow in their proper sequence. The arranging the pages of type for this purpose is called *imposing*, and this term also includes the placing of the furniture between the pages so as to make the necessary margin.

The sizes that may be advantageously worked in the folio foolscap press are as follows, one page of folio foolscap, measuring about 11 by 6 inches in the type. Two pages of quarto foolscap, 7 by 5, or two pages of octavo demy, about 7 by 4. Four pages of octavo foolscap, 5 by 3, or four pages of 16mo. demy, $4\frac{1}{2}$ by 3 inches. Other sizes may be worked, but the above will be found the most generally convenient.

Each page of folio foolscap may be set up and printed separately if desired, and therefore no imposing is necessary in this case. It will however be found more convenient to set up two pages, and to print the second page immediately after the first, whilst the paper is still damp, otherwise it may be necessary to wet the paper a second time, which is not desirable.

Two pages of quarto foolscap, or octavo demy, may be printed on one leaf, by imposing the two pages in one form, and using paper of double the size. The one impression will then produce the first page of one leaf, and the second page of the other; and when the sheet is turned over end for end, or the reiteration is worked, each leaf will be completed. If however, four pages of quarto foolscap, or octavo demy, have to be printed on two sides of a half-sheet, it will be necessary to impose them in two forms, as shown beneath.

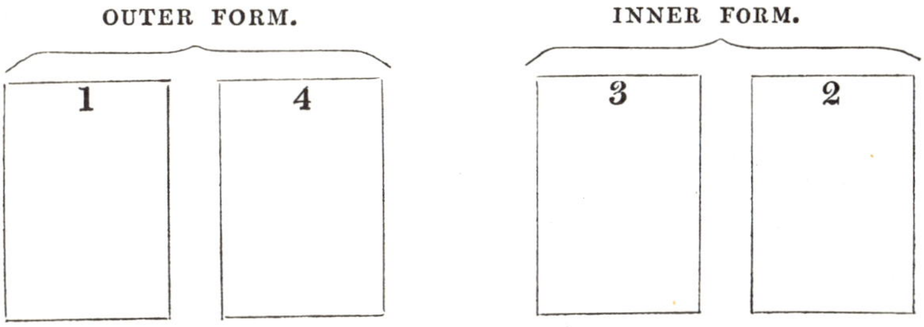

When the sheet is printed and folded, page 2 will be on the back of 1, and page 4 on the back of 3.

Four pages of octavo, foolscap, or 16mo demy, to form two leaves, may be arranged as shown beneath, and two impressions worked together, both sides of the paper being printed from the same form, so that pages 2 and 3 back 1 and 4.

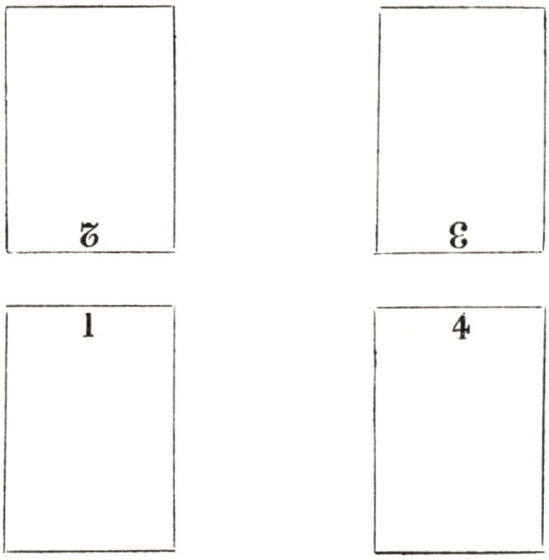

If however eight pages have to be printed upon the half-sheet of foolscap or the quarto sheet of demy, they should be imposed in two forms as shown beneath, and printed at twice, when page 2 will come on the back of page 1, and so with the rest.

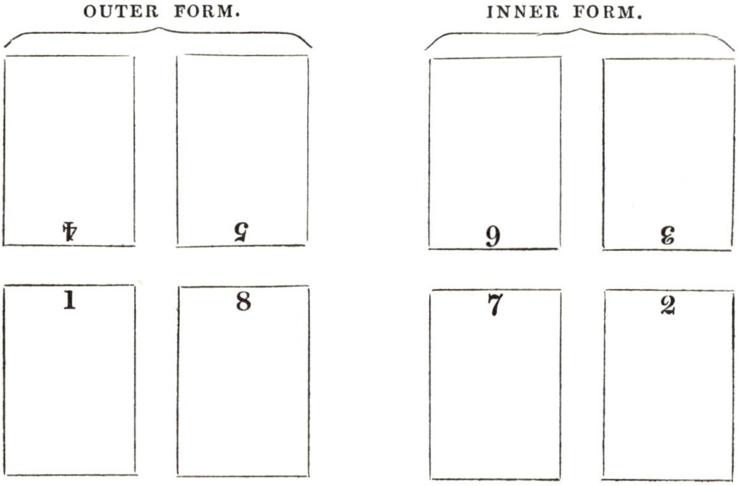

By the modes above described any number of pages that can

be printed in the press may be imposed so as to be worked off correctly. For security however, it is in all cases advisable to fold the first printed sheet to ascertain whether the pages have been properly imposed, before working off the bulk of the impressions.

In dressing the chases for printing two or four pages, it is of importance that the same space be kept between the pages throughout the work, otherwise, when the various sheets are printed they will have different widths of margin; to avoid this very objectionable result, it is usual carefully to proportion the margin in the first form intended for a book, and to use the *same pieces of furniture* for all the rest of the forms.

In imposing four pages in one chase, it will be found desirable to have chases with crosses, which form a separate compartment for every page.

The printing of the paper, and subsequent operations, are the same with both the foolscap and parlour apparatus, and have been already described at pages 24 to 35.

Printing in Two Colours.

It is sometimes desired to print in two or more colours on one page, for instance, with red and black lines alternately; the two colours require to be printed at two impressions, which fit each other accurately. Frequently for this purpose the type is set up in two chases, one for each colour; but great care is then requisite to make the second impression fall exactly in true position on the first; and the amateur will be more certain of success by pursuing the first method, which will be found sufficient for those cases in which one or two lines of small type have to be printed in ink of a different colour from that of the body of the page.

The whole of the type should be set in one chase as for ordinary work, and when corrected, the one or two lines to be printed red should be taken out, and the vacancy filled up with quadrats of the same body as the type; this leaves the lines blank, and in this condition the whole of the sheets should be printed with the black lines; register points being employed as a preparation for the subsequent part of the work to be printed red.

The form should be now carefully washed, the quadrats previously substituted for the lines of words to be printed red should be taken out, and a piece of space-line lead should be cut with a penknife, to the exact length and width of these lines, and placed at the bottom of the vacancy left by the removal of the quadrats, and lastly the removed types should be pushed down upon the *underlay* or the space-line lead; the types to be printed red, will thus be raised above the general surface, as much as the thickness of the space-line lead beneath.

The frisket is now to be covered with strong paper, and an impression taken upon it with red ink from the raised types, and which red impression is to be carefully cut out from the frisket-sheet; the half-worked impressions, being now placed one at a time upon the register-points, may then have the red lines printed upon them with certainty of success.

To enable the amateur to replace the form, after it has been cleaned, to the precise spot it occupied when the black lines were printed, it will be found desirable to mark with a pencil on the bed of the press, a line to show the exact position the form occupied thereupon; this of course should be done before the form is released.

The removal of the form for cleaning might be altogether avoided if the types near to those to be printed with red ink, were covered with a piece of parchment having an aperture of the exact size necessary to expose the red line of types to the action of the roller, the parchment being secured to the wooden furniture either with tacks or paste, and the frisket sheet would prevent the red ink taken up by the parchment from being transferred to the printed impression.

When however the quantities of the work to be respectively printed black and red are more nearly equal, it will be better to print the work at two impressions from the dissected form, without the employment of space-line leads beneath the types.

Thus, beginning with the red impression, let the black lines be all removed, quadrats from the same fount being substituted for them, and the red impression is then first to be worked with register points. For the second impression, the quadrats are removed and replaced by the types laid aside for the black lines, and the types just previously used for red are then to be removed, other quadrats being as before substituted for them.

In this as well as in the last case it is to be observed, that the positions of the chase upon the bed of the press, and also of the register points, should neither of them be shifted in the least between the intermediate stages of the work, or the difficulty will be greatly and unnecessarily enhanced; and it is also desirable that separate rollers should be always used for the two kinds of ink, or otherwise that the roller should be most carefully cleaned between the two processes, either with water or turpentine.

Printing in Bronzes and Gold Leaf.

Printing in gold, silver, and copper bronzes may be executed by the amateur in the following manner. The form should be prepared exactly as for printing with the usual ink, except that to remove all trace of the printing ink, it is desirable that after the corrections have been made, the form should be thoroughly cleansed with lye, and this again well washed away with water, as explained under the head "cleaning the type," page 32.

A small quantity of the *best japanner's gold size* is then to be spread upon a clean distributing plate, with a clean roller, exactly the same as in printing with ordinary ink, and the types are to be "rolled" and the impression taken with the gold size, just after the usual method, but care must be taken in raising the sheet of paper from the types, to prevent any part of the surface from being torn, as the gold size is much more adhesive than printing ink.

The bronze powder is then applied to each impression as it is taken from the press; this is usually done by means of a soft puff about 1¼ inch diameter, made of wash leather stuffed with cotton wool. The puff is gently dipped into a small heap of the bronze to pick up a thin coating of the powder, which is transferred to the impression by gently dabbing the puff in lines over the paper from side to side, or in the direction of the lines of type, commencing at the top of the page and proceeding gradually downwards, care being taken not to smear the gold size. Another impression is then proceeded with in like manner, and so with the remainder until the requisite number is completed; after which the printing should be set to dry for about 24 hours, when the few surplus particles of bronze that remain may be wiped off with a small piece of cotton wool. One coating of the bronze powder upon the puff usually serves for five or six impressions, which if as large as 20 inches square require a ball or puff of 3 inches diameter.

Should it be desired to introduce one or more lines of a different coloured bronze, it may be effected by carefully using a separate puff for each bronze, but of so small a size as not to endanger touching the neighbouring lines.

When the bronzed impression has been dried the work will however present a dull appearance, and the concluding process is to glaze or burnish the printing to give it a solid and metallic lustre. On the large scale, this if effected by placing the face of the impression upon a highly polished plate of steel, a sheet of glazed milled board is laid on the back of the paper, and all three are passed through a copper-plate printing press; the intense pressure to which the bronze is thus subjected, in contact with the polished steel plate, suffices to glaze or burnish the printing, which is then complete.

On the small scale the burnishing may be done either with an ordinary agate burnisher, or with a small flat piece of any hard and smooth stone, the edges of which have been rounded to prevent the paper from being torn. The burnishing will be readily performed by laying the sheet of paper upon a smooth surface of metal, stone, glass, or even on thick Bristol card-board, and traversing the burnisher with moderate pressure over the face of the printing, until it assumes the desired brightness.

Printing may also be executed with gold or silver leaf, which does not require burnishing, and is besides brighter and more solid in appearance than the bronze, but is at the same time more expensive. In printing with the leaf gold, the impression is taken with gold size just the same as for bronze; the leaf gold is then carefully laid on in the ordinary manner of gilding, to cover the whole of the printing; when dry the waste is lightly rubbed off with a small piece of wool, and the paper only requires a gentle pressing to remove the indentations caused by the types.

It is to be observed that in printing either with gold leaf or bronzes, it is indispensable to use enamelled paper or card, or else paper that is very highly glazed, all of which must be used *quite dry*, whether for gold size or common ink. Should ordinary printing papers be used, the bronze or gold would fill up the interstices in the rough surface of the paper, and the whole would appear imperfect and confused, besides using a much larger quantity of bronze.*

---

* The above particulars on printing in gold and bronzes were kindly furnished by Mr. J. Brimmer.

PRINTING MUSIC FROM MOVEABLE TYPES.

Many amateurs have expressed a wish to be provided with the means of printing a small page of music in the parlour printing press; as the mode of printing music from pewter plates indented by punches, is not suitable to the wants of amateurs. Holtzapffel and Co. have therefore provided cases containing a small supply of music type; the cases are made of deal, and are similar in size and appearance to those usually furnished with the parlour press, and described at page 11.

The music cases have each four drawers, and which are divided into 200 compartments for receiving that number of varieties of the characters required in printing music by means of moveable types. An entire fount of music type comprises about 260 characters, but the selection contained in the cases will be found sufficient for most purposes, and should any other sorts be required they will seldom exceed 10 or 12 kinds. From the comparative expense of music type, and with the view of keeping the cost of the first supply of these as moderate as is consistent with utility, only a small number of any one character is supplied; the fount however contains in all about 2800 types, and will in general be found sufficient for printing an octavo page of demy paper, but of course the quantity of music type can be increased to any required extent.

The specimen at the head of this article shows the size and appearance of the music type when set up or combined, and which size has been selected partly with the view of using the pica type supplied in the ordinary cases, for printing the words of the music.

The following specimen shows on the left a bar containing four crotchets printed in the usual manner, and on the right the various characters requisite to compose this one bar are shown separated, so as to explain the mode of combination, and which will be better understood by an inspection of the specimen than by any lengthened description; it may however be observed that the five lines of the stave are all set up simultaneously, and that the mode of combination is highly ingenious and admits of some choice of method in the composition, which gives considerable interest to the above process. Types are also cut for the old Catholic music with square notes, which is still used in some of our cathedral choirs, and has been printed in the little apparatus these pages are intended to explain.

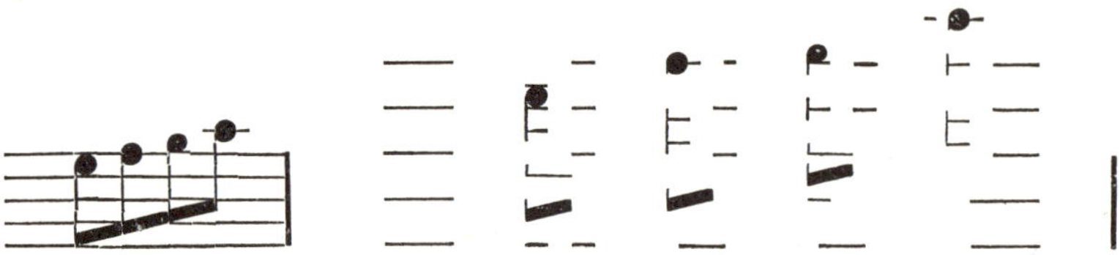

Casting Composition Rollers for Printing.

When a new roller becomes requisite, the amateur will find it most advisable to purchase one ready for use, provided it can be obtained with facility, as the preparations and inconvenience of making rollers will more than equal the expense of purchasing several of them. It sometimes however happens that from locality, a new roller cannot be purchased at all, and it then becomes necessary for the amateur to make them for his own use, in the mode now to be described:—

Rollers for printing are made of a composition of treacle and glue melted together, and cast in a mould around a central wooden core about one inch less in diameter than the roller, so that the composition is about half an inch thick whatever may be the size of the roller.

The mould for the roller should be a perfectly cylindrical tube of pewter, brass, or other metal, of the required diameter, and about half as long again as the intended roller; it should have a smooth internal surface and be uniformly oiled with either neatsfoot oil or with sweet oil. The mould should also be fitted to a stand to hold it steady and prevent leakage; a circular disk being provided to form the bottom of the tube, with a fixed central wire to keep the wooden core in its proper position in the mould. The core should be of the same length as the roller, and when it has been placed around the central wire of the mould, a triangular plate of metal fitting the tube and having a central hole, should be passed over the wire to fix it quite centrally, and also to denote the length of the roller when cast: a weight of some kind should in addition be slid over the wire to keep the wooden core from floating in the composition, the preparation of which will be next described.

The composition of which the rollers are made consists of glue and treacle in the proportions of about 1¼ lb. of glue added to 2 lbs. of treacle, although the exact proportions depend somewhat on the state of the atmosphere. Sometimes a little Paris white is added; this however is not essential, and being of a poisonous nature it is not to be recommended.

The glue should first be broken into pieces not exceeding about one inch square, put into a pan or bason and barely covered with cold water; it should then be allowed to soak for a few hours. The glue is next to be melted in an ordinary carpenter's glue kettle, which is a double vessel, the outer one to contain water, the inner to contain the glue; this arrangement prevents the composition from being overheated and burnt. If a glue kettle is not at hand, a small saucepan supported inside by a larger one containing water will answer the same purpose.

When the glue is completely melted, the treacle should be added, and the whole kept gently simmering over the fire for one or two hours, that the composition may become thoroughly incorporated; to facilitate which, the whole should be frequently but gently stirred with a stick. The composition should be kept on the fire until the water in which the glue was soaked has been evaporated, and the composition has become of a nearly black colour, perfectly uniform, and of such a consistency that it will just admit of being poured in a small stream without forming lumps; the usual test for this is to lift the stick out of the composition, and if the latter just hangs to the stick sufficiently to draw into threads without becoming ropy it is considered to be in the right condition.

If the boiling is continued too long, the composition will become a thick glutinous mass that does not easily admit of being again softened; and therefore if this condition be approached a little water may be sparingly added. Should the stirring or boiling be carried on too actively, air bubbles would be mingled with the composition, and these would make the surface of the roller full of small holes, which are very objectionable.

When the composition has been thoroughly melted, it should be taken off, but allowed to stand near the fire for half an hour, that the impurities and air bubbles may rise to the surface; these may then be skimmed off, and the composition is ready to be poured into the moulds, which should have been previously prepared, and are then quite filled, in order that the upper part of the composition, which almost always contains air bubbles, may be cut off, still leaving the sound roller below of the full measure required, that it may fill out the length of its frame.

After the roller has been poured 10 or 12 hours it may be removed from the mould; this must be done very carefully, rather allowing the roller to slide out by its own weight than by any force applied to it. If the roller should not readily slide out, it shows either that the internal surface of the mould is not perfectly cylindrical and sound, or that it was not sufficiently oiled. If either of these faults exist in the mould, the roller will hang, and in applying force to remove it, its surface will probably be torn, and the roller rendered useless; care is therefore necessary to ensure that the surface of the mould is cylindrical, smooth, and uniformly oiled. When the roller has been removed from the mould it may be cut off to the length, which will be indicated by the points of the triangular plate placed on the core; the cutting will be best effected by wrapping a piece of fine twine or catgut around the roller, and drawing the string tight; this will cut through the composition very cleanly.

Old rollers may be remelted to make new, but this will be found rather more difficult to manage. The old roller should be washed, cut into small pieces, and soaked in cold water; and some new composition should be added with rather a full proportion of glue. But although manufacturers, from economical motives, always mix a large proportion of the old composition with some of the new, and which answers perfectly on a large scale, the amateur will find it easier and better to use new composition every time. In a warm and moist atmosphere the first-named proportions of treacle and glue will make the roller rather too soft, and in a cold dry atmosphere it will be rather too hard. If therefore in practice the roller is found too soft, a little more glue should be added, and if too hard, a little less, as by altering the proportion of glue, the composition may be made of any required consistency.

The above particulars relate entirely to *rollers*, which for ordinary use will be found far superior to the '*balls*' or '*dabbers*' formerly employed. If however, the amateur is so situated that he can neither procure a roller nor a mould in which to make one, a pair of composition balls may be produced without any mould as follows:

Some composition should be mixed, as above described, taking care that it is quite free from lumps or air bubbles; a piece of coarse cloth or canvas may be then stretched on a horizontal board, and some melted composition gently poured into the middle of it, this will gradually extend itself according to the quantity poured on, in a circle to any size that may be required, but about 6 inches diamater, and ¼ inch thick will be sufficient for the parlour or folio foolscap presses. When the composition is cooled it is to be turned over, some wool is put upon the cloth, which is gathered up at the edges and tied or nailed to a central handle.

Section III.—CASES FOR TYPE.

It frequently happens that amateurs possessed of either the parlour, or folio foolscap apparatus, wish to extend their selection of types by the addition of small numbers of different kinds, for instance, small type for foot notes, Italics for ordinary composition, or ornamental types to give variety in headings, title pages, or other purposes requiring embellishment.

To provide for the orderly arrangement of these various sorts, in a compact form and at a sufficiently small cost, three additional kinds of cases are provided that vary in capacity according to the purposes for which they are intended, the smallest being suitable to a very few types, the largest capable of holding a tolerably large supply. These cases will be now described—

Small Type Tray.

This is shown in fig. 9 and measures 11 inches by 5 inches,

it is made of mahogany and inclosed in a case with cover; this type tray has 65 circular compartments which are sufficient to accommodate a small fount of 600 Bourgeois types, which constitute a fount suitable for foot notes, or it will contain about 300

Fig. 9.

pica Italic types, a number that is generally sufficient for use in the parlour printing press. Should it be wished, a proportionate number of any of the smaller sizes of types may of course be substituted.

Fig. 10.

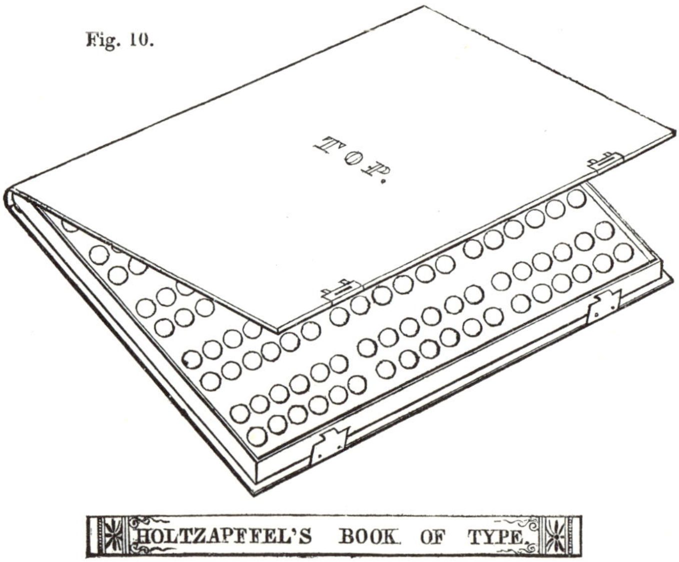

Large Type Book.

The Type Book fig. 10 measures 15 inches by 11. This in external appearance nearly resembles a book and has two clasps to secure the lid, which is padded to keep the types in their cells. The type book has 180 circular compartments, and is usually furnished with a selection of about 1500 types, including eight varieties, with their requisite stops, spaces, and quadrats, namely:

Of Small Pica Roman, No. 15, large capitals and small capitals.
Of Bourgeois Roman, No. 17, large capitals, small capitals and lower case letters.
Of Small Pica Black letter, No. 22, large capitals and lower case.
Of Bourgeois Antique, No. 23, large capitals.

The type book when furnished with the above selection of type, will be found a very useful addition to the parlour printing apparatus, and the selection may be increased in number or varied in kind.

Large Type Trays.

Large type trays measuring 22 inches by 14, exactly resemble the drawers shown in fig. 7, page 38, except that they have no knobs, in fact the type trays form the drawers of the large type case; the boards that constitute the bottoms of these trays project as fillets, so that any number of these trays may be afterwards fitted into a case with appropriate grooves should it be desired.

When the type trays are supplied without a case, to effect their orderly arrangement, a hole is bored through the centre of the end of every tray, and the upper parts of the holes are fitted with dowels or projecting pins, which fit into the lower part of the holes in the tray next above; they will thus fit one upon another in a regular tier, and when furnished with a lid may be secured together by passing two straps around the whole. Two large trays will hold a complete fount as explained at page 36-7, and one tray will suffice for a moderate quantity of capitals and lower case, or for two sets of capitals. The large type trays are found by far the most suitable for additions of type to the folio foolscap apparatus.

Hand Chase.

The hand chase shown in the cut on page 56 is intended to hold a few lines of types, not exceeding altogether three inches in length or one inch in breadth, and for all purposes within that size it will be found very efficacious. As for instance the printing of small labels, the headings of account and other books; the stamping or endorsement of various business papers, such as bills of exchange, script, coupons, &c.; the printing of the titles of drawings or maps; marking linen, directing parcels, and numerous other cases, where it is desired to print a few explanatory words, but to which several purposes either from the large size of the paper or other circumstances the ordinary printing press is inapplicable.

Fig. 11.

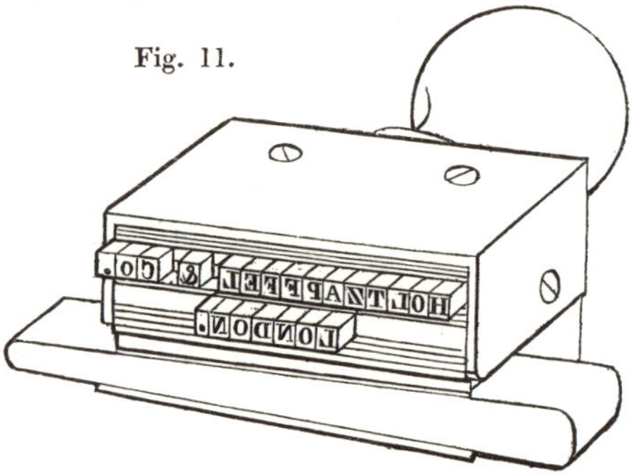

The hand chase is made to suit the length of the three inch leads, some of which are supplied with it, and in use the types are set up exactly the same as in the printer's composing stick, filling each line rather tightly, and when the whole of the type has been set up, it is locked securely in the chase with the wedge as shown in the wood-cut. The hand chase is fitted in a painted case 13 inches by 5½ containing in addition a 3-inch inking roller, a box of ink, an inking tray, and a cushion upon which to lay the paper to be printed.

When the type has been set up, the ink should be distributed on the tray as explained at page 26, and the type inked with the roller, the chase being held in one hand, the roller in the other.

The cushion should next be slid under the part of the sheet to be printed, and the hand chase placed steadily in its proper position, the pressure should then be gradually applied, observing that no stamping blow should be given, on the contrary, that the weight of the body should be quietly thrown on the chase, and that, without any direct muscular exertion on the part of the individual. It will be found that impressions may be produced in this manner equal to those of ordinary printing, especially if the paper is slightly damped in any way; a ready mode of doing which is to place a few thicknesses of moistened blotting paper above and below the part to be printed upon.

The small type tray, fig. 9, is usually supplied along with the hand chase, and will be found to contain a sufficient supply of one fount of type for ordinary purposes, and where ornament is desired any selection of type may be made.

Monotype Printing Press.

Some time prior to the invention of Mr. Cowper's ingenious parlour printing press, previously described, Charles Holtzapffel contrived the Monotype press for printing labels, or short descriptions, of which from their nature *but one single copy is required,* as the labels for specimens of natural history, works of art, &c., which objects commonly differ in almost every respect, as to character, structure, or locality.

The principal inducement for the trouble of arranging types in a chase, as in ordinary printing, namely the *facility of afterwards multiplying copies,* does not exist in this case, and in consequence, the labour of printing would be rarely undertaken for such single copies.

By the arrangement of the monotype press however, the trouble of the process is on the whole much diminished, as the types or letters, are only picked out of the trays one at a time, and printed from separately by means of a blackened paper, the types are immediately returned to their places in the tray, and the entire process is then accomplished, and that with considerable facility of manipulation. As in common printing, the perfection of the result is mainly dependent upon the accurate structure and uniformity of the types, which has been explained at page 13; we will now proceed to describe the monotype press represented in fig. 12, over leaf.

The Monotype press consists of a small casting, in iron, with a little sliding table A, which is guided in a straight line, by the ledge or rib on the principal piece, and is fixed to the latter by the little binding screws E, E, (one only being lettered,) the card or paper to be printed upon, with a piece of the prepared blackened paper laid above it, is attached to the table A, by means of the little clamps and screws upon the surface of the table.

Suppose it to be desired to print the word ASBESTOS, the letter A is held in the notch at the end of the arm B, with the thumb and fore finger of the left hand, with its face downwards and the nick of the type outwards or towards the operator, and the type is then pressed down by means of the lever C, with the right hand. The small screw in the lever is adjusted once for all to touch the arm A, so as to prevent the letters and stops from being pressed so hard upon the paper as to indent its surface. When the first letter has been printed, it is necessary to move the table and paper to the left hand, as otherwise the following letter of the word would be impressed upon the same spot as the preceding one.

Fig. 12.

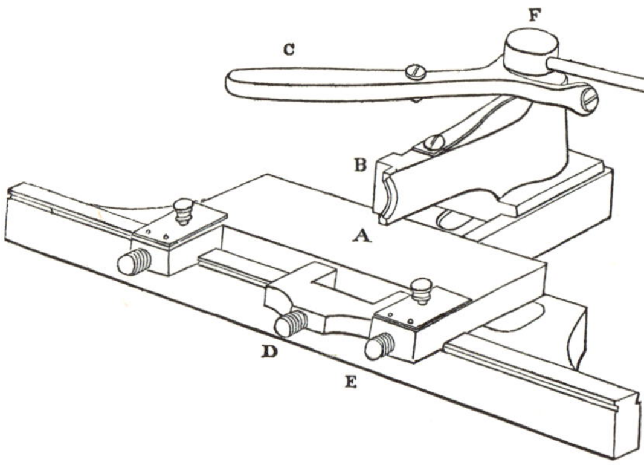

The quantity the table must be shifted will be given by the shaft of the type itself, as the width of the shaft of every letter has been fixed by the typefounder, so as to be exactly proportioned to the sizes of the respective letters. To proceed, the type for the next letter of the word ASBESTOS, namely, S is placed *with the nick outwards*, between the arm E of the table and the loose, stop D; (in fact the type is as it were pinched between them in the direction of its width,) the stop D, is then fixed; the table A, is loosened and re-fixed in *contact* with D, and it will then have been shifted, exactly the space necessary for the letter S, or the precise width of the particular type, and lastly the letter S is to be impressed on the paper and the type returned to the tray. The same routine is gone through with the succeeding letter B, namely, the stop is shifted exactly the width of the B, and the table is run up to its new position, and then B the third letter of the word is impressed, and replaced in its cell; after a little practice, this routine will be found by no means a slow operation.

It should have been observed, that one of the screws E, (not lettered,) is purposely left stiff; it is adjusted to allow the table to slide freely without shake, and is not afterwards disturbed.

Some prefer holding the type in the nick B, with the thumb only, placing the fore finger at the back of the sliding arm.

In common printing, *spaces* and *quadrats*, or blank letters, are placed between every word for their separation; it is best to take up the quadrat and the first letter of the following word *together*, otherwise two adjustments of the table would be necessary. T o s e p a r a t e t h e l e t t e r s t h u s, the types and the particular spaces should be used together as above.*

One line having been completed, the apparatus is shifted for a new line by releasing the screw F, and advancing the arm B. To save the expense of a stop similar to D for that purpose, and in fact as the most convenient arrangement, the face of the instrument is graduated; three of the divisions will be found equal to the height of the small type, called Pica, and four to that of the large, or Great Primer, but 5, 6, 7, 8, or any other number of divisions may be used, for giving more space between the lines; which purpose would be effected in ordinary printing by inserting leads between the lines of type as already noticed.

As the fount of type required for the Monotype press is only a single letter of each kind, the variety of sorts may be extended without a great increase of expense; six alphabets, &c., accompanying the press, namely,—

Great Primer, A A a

Pica, A A a

And a duplicate set of each is also added, in case of accident; making a total of 472 types.

The blackened paper which accompanies the press, called also camp paper, is prepared for the purpose of writing by means of a style or blunt point, in the method familiarly known as manifold writing, in which the original letter and the copy are written at the one operation; or even two, three, or more letters may be written at the same time. The camp paper may be made by grinding together lamp black, and any unctuous substance that will not dry, such as neat's-foot oil, sperm oil, lard, soap, &c.; the first seems to answer best. The pigment, which is laid on the one or both sides of thin hard paper with a sponge or brush, is allowed to remain on for a few hours, and the

* This method of distinguishing particular words is called German Italics, and is adverted to on page 63.

greater part is then rubbed off. Colours may be employed instead of lamp black.

In printing upon paper, it is necessary to put two or three thicknesses of waste paper or card beneath, to serve the purpose of the blanket in the common press; card does not require anything beneath it. Two or even three impressions may be made at once, by placing alternately the white and the prepared papers; two copies succeed very fairly, especially when the lower is on card, and of course, the thinner the papers, the greater will be the distinctness of the impressions.

The Monotype press may also be used with printing ink, but in that case its application is limited to one impression at a time, and it becomes necessary to have an inking cushion and a distributing roller, as in the method of printing explained in the following article, and which proceeding is accomplished without a press of any kind.

Professor Willis's Mode of Printing Architectural Plans, &c.

▷ A very useful application of Monotype Printing has been very recently introduced by Professor Willis, of Cambridge—indeed whilst these papers have been passing through the press; namely, its employment in the production of diagrams used in the illustration of architectural science; but the mode is also more or less applicable to diagrams and objects of other kinds.

It is well known that on looking at the ground plans of architectural buildings, the majority of them will be found to be either parallelograms, as in ancient temples and numerous other buildings; or in the form of a cross, as in the ground plans of most cathedrals and churches; and that in all cases the walls consist of various rectilinear parts, occasionally mingled with parts of circular outline. And further, that the walls are frequently surrounded externally or internally with rows of columns, pilasters, or buttresses, arranged in a variety of ways, the columns being commonly circular, sometimes semicircular, or rectangular as in pilasters, and that various other mixed and complicated forms occur in architectural plans, as in the piers and clustered columns of gothic architecture.

It becomes a very lengthy and tedious business to draw these numerous columns and details individually, and then to fill them with black or colours, with a due regard to their equality of size and form; but the matter admits of very easy and rapid execution when the typographic art is called in to assist.

Suppose, for example, it is desired to represent a Grecian temple surrounded by one or two rows of columns; the general outline of the building is first drawn in pencil, with additional cross lines, the intersections of which denote the exact centres or positions of all the columns. A type is next obtained of the exact diameter of the shaft of the columns, as for instance, the uncut extremity of an ordinary drawing pencil, if its size should happen to correspond; the type is then moistened with printing ink, and impressed upon the intersecting lines that denote the position, say of the column at the one angle of the building, and the wooden type is alternately inked and pressed upon the drawing at every succeeding intersection; but which process, after one or two trials, may be done with considerable facility, and is obviously far more rapid and exact than first drawing the outlines and then filling in the spaces with Indian ink or colours.

The saving of time is of course the greater the more complex the section of the column; but whether the section is semicircular as in half columns, rectangular for pilasters, or of the curious mixed outlines met with in gothic columns and buttresses, it is only necessary to shape the flat-ended stick or piece of wood with the penknife, or still better with the chisels, gouges, and files used by the cabinet-maker, and the type once prepared, the monotype method reduces the most complex forms to the same facility of execution as all others.

The method is equally applicable to printing the plans of the walls; suppose the latter should be $\frac{1}{4}$ inch thick on the drawing, two or three pieces of wood, of one quarter of an inch thick, one inch wide, and of various lengths, are provided. The edge of the slip of wood is inked, and pressed on the paper; of these slips one is wanted as short as the shortest piece of wall shown on the plan, one as long as can be conveniently used, and one or two of intermediate lengths, as then any arbitrary length of wall may be printed on the plan, provided the wall type is shorter than the entire length of such piece, but longer than the half of the same; the type is placed first to the one extremity of the wall, and then to the other extremity, and therefore some portion of the central part will have received two impressions, which is immaterial. In this manner the apertures for doors and windows may be avoided, and the precise terminations of internal and external angles, may be impressed; and in a similar manner, if the wall should be at places thicker than the type, these may be completed at two or more impressions, varied laterally.

It only remains to be observed, that for the convenient inking of the types, it is best to employ a composition cushion, namely a layer of the glue and treacle cast upon a flat board; the ink is first spread or distributed with a roller upon the cushion, and the types are then applied at various parts of its surface, before a re-distribution of the ink upon the cushion becomes necessary.

When certain parts are distinguished from the rest by colours, such as red, blue, green, brown, &c., a roller and cushion must be provided for every colour. The office of the blanket used by printers, will be rendered by a few sheets of soft paper, or a stout table-cover, placed beneath the drawing.

The same method of printing may be used for representing various other forms required in diagrams for lectures, such as the teeth of wheels and racks, threads of screws, straight, circular, and dotted lines; and with the further assistance of a few printer's types, all the letters of reference and inscriptions may be printed on diagrams, railway plans and sections, and even on finished drawings, with great distinctness, by this useful modification of typographic printing.

Section IV.—SPECIMENS OF TYPES, &c.

It will be imagined, that these specimens introduced in the concluding section of this pamphlet, are quite insignificant in point of number compared with those types which are more or less used by printers; as some of the enormous types for posting bills are two or three feet high, or nominally above 200 lines pica, (see foot of page 14,) and others called *diamond* types are so minute as only to be legible to those possessed of good visual organs.

Of the plain types various intermediate sizes, as well as those larger and smaller than the specimens, will be supplied to order; and the same may be also partially said of the ornamental types, which, however, are far less numerous in their varieties of size.

No. 1. Five Line Pica.—Price of Capitals & Figures, 6s. 6d. the dozen.

Roman
TYPE.

No. 2. Five Line Pica.—Price of Small letters, points, and spaces, 4s. 6d. the dozen.

Italic
TYPE

The distinction conferred on certain words of a sentance *by printing them n Italics,* may be attained by the method sometimes called G e r m a n t a l i c s , that is by the separation of the letters with thin spaces.

The mode of separation is almost imperative in German typography, in vhich the **German Text Character,** called in England **B l a c k L e t t e r ,** is almost always used—and which mode of separation although erhaps inelegant, may be sometimes convenient to the English amateur, as t may obviate the necessity of keeping both the Roman and *Italic* types.

No. 3. Canon types—Price of capitals and figures. 4s. 0d. the dozen.

The Atlantic OCEAN.

No. 4. Canon types—Price of small letters, points and spaces, 2s. 6d. the dozen.

The Pacific OCEAN.

No. 5. Two Line English types—Price of capitals and figures, 2s. 0d. the dozen.

The INDIAN Ocean, the Mediteranean, the Black and Red Seas.

No. 6. Two Line English types—Price of small letters points &c. 1s. 3d. the dozen.

The LAKES Superior, Michigan, Erie, Huron & Ontario.

No. 7. Two Line Small Pica types — Price of average sorts, 9d. the dozen.

EUROPE ASIA AFRICA and AMERICA, constitute the four great quarters or divisions of the GLOBE.

No. 8. Two Line Small Pica types — Price of average sorts, 9d. the dozen.

AUSTRALIA has been recently considered as the fifth quarter or division of the habitable World.

No. 9. Great Primer types — Price of average sorts, 6d. the dozen.

The River AMAZON, usually considered the largest river in the World, runs almost across the broadest part of S. AMERICA, and receives the waters of 200 various tributary streams.

No. 10. Great Primer types — Price of average sorts 6d. the dozen.

The rivers Missisipi, Plata, Orinoco & St. Lawrence — Nile & Niger — Lena, Ganges & Euphrates — Volga, Danube, Dneiper, Don & Rhine, are severally remarkable for their length.

No. 11. English types — Price of average sorts 4½d. the dozen.

ETNA, a celebrated burning mountain of SICILY, which is called by the natives *Monte Gibello.* Pliny who lived 435 years before Christ calls it the Pillar of Heaven, on account of its great height, which is now generally reckoned to be about 11.000 feet; and its circumference at the base 70 miles.

No. 12. English types — Price of average sorts 4½d. the dozen.

VESUVIUS, a mountain of Italy, about 8 *miles S. S. E. of Naples, celebrated for its volcanic eruptions. It rises in a gentle swell from the Bay of Naples, to the height of* 3731 *feet. — Although the crater is nearly a mile and a half in circumference, it only measures* 350 *feet in depth from the ridge.*

No. 13. Pica types — Price of average sorts, 3d. the dozen.

STROMBOLI, the most northerly of the LIPARI ISLES, is about 10 miles in circumference, and has from time immemorial been remarkable from its possessing perhaps the only volcano whose fires are in a constant state of activity. The eruptions last for a few moments at a time, and recurring at short intervals, have procured for Stromboli, the name of the great lighthouse of the Mediterranean.

No. 14. Pica types — Price of average sorts, 3d. the dozen.

Mount HECLA is a noted volcanic mountain in the southern part of Iceland. It is divided into three points at the top, the highest of which, in the middle is computed to be 5000 *feet above the level of the sea. Hot vapours issue from several small openings near the summit, and while the thermometer stands in the air below freezing, when set on the ground it rises to* 120 *and even* 150 *degrees.*

No. 15. Small Pica types — Price of average sorts, 3d. the dozen.

HERCULANEUM, an ancient city of ITALY, about 5 miles E. by S. of Naples, near where the modern Portici now stands. It was destroyed by an eruption of Mount Vesuvius in A. D. 79; and its site was long a matter of doubtful discussion, it having been completely buried under volcanic substances to the depth of 70 feet; Pompeii was also overwhelmed by the same eruption.

No. 16. Small Pica types — Price of average sorts, 3d. the dozen.

LISBON was partially shaken by an Earthquake in 1531. *but on the* 1*st. of November* 1735, *it experienced the most severe recorded in modern times. The shock was so violent that almost all the public buildings, and* 6000 *houses were overturned; the loss of lives was estimated at* 30,000. *The conflagration that immediately followed was even more destructive to this city than the earthquake.*

No. 17 Bourgeois types — Price of average sorts, 3d. the dozen.

GEYSER, the name of some remarkable intermitting springs of hot water, in ICELAND. The most magnificent of these are called the GREAT and the NEW GEYSER, situated about 78 miles north of SKALHOLT. They throw up large jets of water to the height of 90 or 100 feet, accompanied by a noise like the firing of cannon, and a trembling of the ground in the neighbourhood, which effects are doubtless attributable to the internal fires of Mount Hecla.

No. 18. Bourgeois types — Price of average sorts, 3d. the dozen.

The FALLS of NIAGARA are divided into two cataracts, that on the north-western or Canadian side measuring 600 *feet wide, and falling without interuption from rocks,* 150 *feet in perpendicular height, that on the American side, measuring* 350 *feet in breadth and* 160 *in height. The spray from the water rises to a great altitude, and in sunshine forms beautiful rainbows; the noise of the falls is sometimes heard the distance of* 40 *or* 50 *miles.*

No. 19. Nonpareil types — Price of average sorts, 3d. the dozen.

PALMYRA the ruins of a magnificent city of ASIA, situated in the center of the desert of SYRIA. It appears to have been founded by Solomon, under the appelation of Tadmor, and from its situation was called Tadmor in the Wilderness; but the present remains being evidently of Grecian origin, cannot be supposed to be those of the city of Soloman. The principal and most entire ruin is that of the Temple of the Sun, and within its spacious court about thirty Arab families, the only present inhabitants, have erected their mud cottages.

No. 20. Nonpareil types — Price of average sorts, 3d. the dozen.

The RUINS of THEBES are the remains of a great city, which at an early period was the capital of Egypt, The ancient structures still remain in a state of wonderful preservation, and there are not to be found in the world any monuments more remarkable both for their stupendous magnitude, and for the high antiquity they claim. The ruins extend about seven or eight miles along the banks of the Nile, almost the whole of which space is covered with magnificent portals, obelisks, columns, colossal statues, and sepulchral monuments.

No. 21. Great Primer Script type.—Price 1s. 0d. the dozen.

Holtzapfel, a town of Germany, in the Duchy of Nassau, situated at the foot of a mountain, on which are the remains of an ancient castle, the original seat of the Princes of Nassau. The population of the town of Holtzapfel is twelve hundred, it is 4 miles N. E. of Nassau, and 10 miles from the Castle of Ehrenbreitstein.

No. 22. Small Pica Black Letter type.—Price 4d. the dozen.

Teneriffe, one of the Canary Isles, is situated off the Coast of Africa; its peak which is above twelve thousand feet high, is a most useful land mark.

No. 23. Bourgeois Antique tyqe.—Price 4d. the dozen.

AMSTERDAM, the Metropolis of Holland, is a large, rich, populous and commercial city, seated at the mouth of the Amstel, where it falls into an arm of the sea, called the Wye. The principal streets are intersected by canals, whence it has been contrasted to Venice.

No. 24. Great Primer Antique type.—Price 8d. the dozen.

MADRID, the most elevated capital in Europe, is upwards of 2200 feet above the level of the sea.

No. 25. Two Line Small Pica Antique type.—Price 1s. 0d. the dozen.

MECCA the birth place of Mahomet

No. 1.

ETNA

No. 3.

ETNA, a cel

No. 5.

ETNA, a celebrated b

No. 7.

ETNA, a celebrated burning

No. 9.

ETNA, a celebrated burning mountai

No. 11.

ETNA, a celebrated burning mountain of SICI

No. 13.

ETNA, a celebrated burning mountain of SICILY, which is

No. 15.

ETNA, a celebrated burning mountain of SICILY, which is called

No. 17.

ETNA, a celebrated burning mountain of SICILY, which is called by the natives

No. 19.

ETNA; a celebratcd burning mountain of SICILY, which is called by the nativcs *Monte Gibello.* Pliny

No. 21.

Etna a celebrated burning mountain of Sicily which is ca

No. 22.

Etna, a celebrated burning mountain of Sicily, which is

No. 23.

ETNA, a celebrated burning mountain of Sicily, which is

No. 24.

ETNA, a celebrated burnin

No. 25.

ETNA a celebrate

Note—Types both larger and smaller than the above specimens will be obtained to order.—The same may be partly said of the Ornamental Types on pages 70 to 73.

No. 26.

EUROPE.

No. 27.

ENGLAND.

No. 28.

SCOTLAND IRELAND

No. 29.

LONDON.

No. 30.

EDINBURGH.

No. 31.

DUBLIN.

No. 32.

WALES.

No. 33.

MANCHESTER.

No. 34.

BRISTOL.

No. 35.

LIVERPOOL.

No. 36.

GLASGOW.

No. 37.

NEWCASTLE.

No. 38.

SHEFFIELD.

No. 39.

BIRMINGHAM.

No. 40.

OXFORD CAMBRIDGE.

No. 41.

MILFORD HAVEN.

No. 42.

PERTH & ABERDEEN.

No. 43.

ISLE OF WIGHT.

No. 44.

HAVRE DE GRACE.

No. 45.

Portsmouth.

No. 46.

PLYMOUTH SOUND.

No. 47.

LONDONDERRY.

No. 48.

THE ISLE OF ANGLESEA.

No. 49.

THE HIMALAYA MOUNTAINS.

No 50.

THAMES, SHANNON & CLYDE.

No. 51.

The Straits of Gibraltar.

No. 52.

THE MEDITERRANEAN SEA.

No. 53.

SARDINIA, SICILY & MALTA.

No. 54.

CANDIA & CYPRUS.

No. 55.

Vienna, Berlin and Munich.

No. 56.

BOHEMIA & HUNGARY.

No. 57.

OVERLAND ROUTE

No. 58.

LONDON, PARIS & MARSEILLES.

No. 59.

ALEXANDRIA, CAIRO, SUEZ & ADEN.

No. 60.

BOMBAY, MADRAS, CALCUTTA & HONG KONG.

No. 61.

Sumatra Java and Borneo.

No. 62.

THE INDIAN ARCHIPELAGO.

No. 63.

AUSTRALIA & NEW ZEALAND.

No. 64.

Holland, Belgium, the Rhine & Switzerland.

No. 65.

THE ALPS & PYRENEES; SPAIN & PORTUGAL.

No. 66.

The Bay of Biscay.

No. 67.

PETERSBURGH, MOSCOW, & ODESSA.

No. 68.

DENMARK, SWEEDEN, NORWAY & GREENLAND.

No. 69.

The North Polar Regions.

No. 70.

BAFFIN'S BAY, CANADA, THE UNITED STATES, WEST INDIA ISLANDS.

No. 71.

Columbia, Peru, Brazils & Cape Horn.

No. 72.

THE SOUTH POLAR REGIONS

No. 73.

No. 74.

No. 75.

No. 76.

No. 77.

No. 78.

No. 79.

No. 80.

No. 81.

No. 82.

No. 83.

No. 84.

No. 85.

No. 86.

No. 87.

No. 88.

No. 89.

No. 90.

No. 91.

ORNAMENTAL BORDERS.

No. 92.

No. 93.

No. 94.

No. 95.

No. 96.

No. 97.

No. 98.

No. 99.

No. 100.

No. 101.

No. 102.

TYPES FOR BORDERS ARE VARIOUSLY COMBINED.

No. 103.

No. 104.

No. 105.

No. 106.

No. 107.

No. 108.

No. 109.

No. 110.

No. 111.

No. 112.

No. 113.

No. 114.

The Dashes 103 to 109 are composed of from 3 to 7 pieces, and admit of numerous interchanges.

No. 115

No. 116.

No. 117.

No. 118.

No. 119.

No. 120.

Various other Ornaments and Royal Arms, in addition to those shown, may be had if required.

PRICE LIST OF HOLTZAPFFEL & CO.'S PRINTING APPARATUS.

SECTION I.—Cowper's Parlour Presses and Apparatus.

	£	s.	d.
COWPER'S PARLOUR PRINTING PRESS, with a galley-chase, a box of ink, a composition inking roller, and a distributing tray	1	14	0
SMALL DEAL TYPE CASE, painted with four drawers; three of them partitioned to contain an assortment of about 2500 types, and a proportionate supply of leads and brass rule; the fourth drawer contains reglet, furniture, side and foot sticks, quoins, &c.	2	16	0
SET OF EXTRAS—comprising transfer composing stick, bodkin, forceps, mallet, shooting-stick, plainer, brush, and turpentine for cleaning the type, two quires of demy printing paper, cut into suitable sizes for the press, and one pair of damping slates.	0	12	0
GALLEY CHASE seven inches square inside	0	4	0
Total amount of the plain Apparatus complete . .	5	6	0
COWPER'S PARLOUR PRINTING PRESS, japanned and finished in the best manner, and fitted with a drawer, in other respects as above	2	2	0
SMALL MAHOGANY TYPE CASE, with brass lock and handles, in other respects as above	4	4	0
SET OF EXTRAS, comprising Transfer Composing-stick, &c., as above	0	12	0
GALLEY-CHASE seven inches square inside	0	4	0
Total amount of the best Apparatus complete . .	7	2	0
DUPLICATE SET OF 2500 TYPES, and which may be contained in either of the above cases	1	12	0

SECTION II.—Folio Foolscap Presses and Apparatus.

	£	s.	d.
FOLIO FOOLSCAP PRINTING PRESS, on the principle of Cowper's Parlour Press, suitable to printing the half sheet of Foolscap, or the quarto sheet of Imperial, external measurement of the press 21 by 11 inches, measurement of the bed 15 by 10 inches, with two iron chases, register points, &c. The press varnished and japanned, complete	4	14	6
LARGE DEAL TYPE CASE, with six drawers, and measuring externally 24 inches by 18, and 11 inches high, with iron handles, lock and key			
Four of the drawers are partitioned after the Printer's method for holding 9000 types of the following varieties.			
Great Primer, Roman, Specimen No. 9; viz., capitals, figures, points, spaces, quadrats, &c.			
Pica, Roman, No. 13; large and small capitals, lower case (small letters), with accented vowels for printing the foreign languages, figures, points, spaces, quadrats, and space line leads, complete.			
Bourgeois Roman, No. 17; capitals, figures, points, spaces, quadrats, &c.			
Bourgeois Antique, No. 23; capitals, figures, points, spaces, quadrats, &c.			
Two of the drawers contain space line leads, furniture, side and foot sticks, quoins, and reglet; also a mallet, shooting stick, planer, bodkin, Printer's composing stick 9½ inches long, brush for cleaning the type, a pair of thick damping slates, &c., proportioned to the size of the Foolscap Press	9	9	0
Six inch composition inking roller in frame and case	0	10	6
Large box of superfine printing ink	0	8	0
Total charge for the Foolscap Press and Apparatus in the less complete form . .	15	2	0
FOLIO FOOLSCAP PRINTING PRESS, exactly like the one last described, but with the following additions, namely, an iron bed half an inch thick, planed quite level and true, to increase the permanent accuracy of the Foolscap Press, and an iron counterpoise, to facilitate the working of the same	7	7	0

SECTION II.—Continued.

	£	s.	d.
Folio Foolscap Printing Press brought forward	7	7	0
LARGE DEAL TYPE CASE with eight drawers, similar to the case with six drawers above described, but three inches higher, and containing a considerably greater supply of each of the kinds of type specified in the foregoing description, together with the addition of Great Primer No. 9. lower case letters, Pica Italic No. 14. capitals, lower case letters, points, and spaces, and Bourgeois Antique No. 23. lower case letters, making the total number of types about 17,000; together with a proportionate increase of space line leads, furniture, &c., and with the addition of 21 pieces of brass rule of three varieties, and all 16 inches long	16	16	0
Six inch composition inking roller in frame and case	0	10	6
Large box of superfine printing ink	0	8	0
Composing frame to receive the drawers of the type case	1	4	0
Inclined galley with moveable bottom	0	18	0
Four extra chases, two of them with crosses	0	10	0
Total charge for the Foolscap Press and Apparatus in the more complete form	27	13	6

SECTION III.—Cases for Additional Types.

	£	s.	d.
SMALL TYPE TRAY, 10 by 6 inches, with a selection of about 600 Roman or Italic types of small size, of either of the numbers 17 to 20	0	15	0
The empty type tray	0	5	0
TYPE BOOK 15 by 12 inches, with a selection of about 1500 types, comprising 8 varieties of small types for headings, cards, &c., as described on page 55	2	2	0
The empty type book	0	12	0
LARGE TYPE TRAY 22 by 24 inches, partitioned after the mode of the printing office, for containing larger quantities of type of any kind; namely, the tray without types	0	7	0
MUSIC TYPE CASE of deal, painted, uniform in size with the Small Deal Type Case described on page 11. The Music Type Case contains four drawers, the whole of which are partitioned to receive an assortment of 2800 music types, of 200 different kinds, as described on page 49. The case with music types complete	5	15	6
HAND CHASE, in a painted case, with cushion, roller, ink, and inking tray	0	15	0
The Hand Chase alone	0	7	6
HOLTZAPFFEL AND CO'S. MONOTYPE PRINTING PRESS, for labels, &c. with a brass table, 4 inches by 2½ inches (see cut page 58) contained in a mahogany case, 11 inches by 9 inches, and six inches deep, with a lock and key; two trays with 120 cells in each, containing the six alphabets of type, &c.; a duplicate or reserve set of type, in all 472 pieces, prepared black paper, and 250 blank cards, as large as the table, which is the limit of size to which this press applies	5	15	0

SECTION IV.—Types of Various Kinds.

(*Types kept in Stock.*)

The plain types kept in Stock, Nos. 1 to 25; pages 63 to 69 are sold in small quantities at the prices per dozen quoted at the head of the specimens; which prices serve for all numbers below 50 dozen. When more than that number of one fount is supplied at the same time, a reduction in the charge is made proportionate to the quantity furnished. When the weight of the types supplied of one fount exceeds ten pounds, they are sold at a further reduction of price, and by the pound weight, instead of by the number of dozens. A few dozens of the smaller sizes of types may, if required, be readily and economically forwarded *by post;* but in this case the lowest price that can be charged for any parcel is *one shilling;* exclusive of the postage.

Ornamental Types procured to order.

The Ornamental Types Nos. 26 to 72 pages 70 to 73 may be considered in general to be about twice the price of the plain types of similar sizes already particularized; but some few of the ornamental types are about four times the price of common types of equal size. It is however, in all cases desirable, that the entire quantity of ornamental types required should be ordered at the same time, as however small the quantity, the lowest price that can be charged for one parcel of any particular fount is two shillings; but more exact details of the prices of these and the other specimens of printing materials will be furnished on specific application.

30*th June*, 1846, 64, *Charing Cross, and* 127, *Long Acre, London.*

NOTES ON THE TEXT

p. 5 *The Parlour printing press*

There are three sizes of press mentioned in the manual, referred to on pp. 7, 8, and 35, and in the Price List, pp. 78–9.

The ordinary size, with a bed 10 inches by 8, was made in two qualities. With chase, ink, roller, and distributing tray, the plain press cost £1 14*s*; the best, with the same equipment, but 'japanned and finished in the best manner, and fitted with a drawer', cost £2 2*s*. It is a surviving example of this latter sort that is described in detail in the introduction, p. xxxi–xl.

The wooden folio-foolscap press, with a bed 15 inches by 10, cost £4 14*s* 6*d*; the same with the addition of a half-inch iron bed, and an iron counterpoise, cost £7 7*s*.

The half-sheet demy, with a bed 20 inches by 12, is only mentioned in the introductory pages of the manual, pp. 7 and 8, and not where one would expect to find it, at the beginning of section II, p. 35, nor in the Price List, pp. 78–9. It is included in the Price List (dated 1873/1897?) pasted into James Mosley's copy of the manual where it is offered, with an iron bed and counterpoise, at £15; its size is there given as 20 inches by 13.

A fourth size of press appears in this list, a broadside foolscap made entirely of iron, with a bed 20 inches by 15, costing £22.

p. 5 *Timothy Bramah*

Actually *Joseph Bramah*. Patent No. 2977, October 15 1806. See illustration, p. x.

p. 6 *is sufficiently near to the hinges*

Should read 'is sufficient near to the hinges'.

p. 8 *in these large presses*

See note to p. 5 on the sizes and prices of the various presses made.

p. 11 2500 *Pica Roman types*

About 9 or 10 lb.

p. 13 *Fig. 3 represents an ordinary type*

The shank should have been described as rectangular, not square. This appears to be the archetypal representation of a type, found in nearly all the later manuals for amateurs. Curiously enough there were no similar drawings in the manuals for professionals published before this, although an unannotated engraving of a type is shown in the article 'Imprimerie en caractères' in the *Encyclopédie* of Diderot and d'Alembert.

p. 13 *The spaces and quadrats*

Mid spaces are not included in this list, nor is space allotted for them in the diagram of the lay of the case on p. 19. They were in general use in the trade at this time however, and are specifically mentioned both in Johnson, 1824, and Hansard, 1825.

p. 13 *and the 4 em quadrats measure about ½ an inch*

A 4 em pica quadrat would measure about ⅔ inch.

p. 13 *the hair space about one tenth the thickness of the em quadrat*

Johnson, 1824, and Hansard, 1825, both quoting from Stower's *Printer's Grammar*, 1808, say merely that they are 'cast remarkably thin'. At the end of the century Southward specifies seven to ten to the em according to the body.

p. 15 *the length of the leads should be exactly equal to that of the line of type*

On the contrary they should be cut very slightly short so as not to prevent the lines of type from being locked up tight. It would have been better to advise the amateur to place an em quadrat at the beginning and end to keep the end types in their proper positions.

p. 16 *Transfer composing stick*

This seems to have been an instrument peculiar to the outfits supplied by Holtzapffel. The method of using it had to be the reverse of that normally employed so that the top of each line of type should be at the open side of the stick and thus easily transferred line by line to the bottom of the galley of type already set. It will be noticed that the transfer composing stick was not adjustable in length; it was presumably set to three inches since the majority of the space line leads supplied with the equipment were this length (see p. 15).

p. 17 *Two quires of demy printing paper*

The octavo size would have been $9 \times 5\frac{1}{2}$ inches; the size of the other quire would presumably have been cut down from demy quarto, 11×9 inches, to the 10×8 which was the largest size the press would take without any overhanging at the sides.

p. 19 *Fig. 4*

No space is allotted to mid spaces. See note to p. 13, *supra*.

p. 19 *The transfer composing stick*

See note to p. 16, *supra*.

p. 20 *taking out all the thick spaces in the line, and substituting thin ones*

The curious lack of mid-spaces now makes itself felt. The rest of the advice on setting is excellent within this limitation.

p. 23 *the quoins are driven in tight, and the planer is also again used*

Very dangerous advice: any letters still high after locking up would only be damaged by being forcibly levelled in this way.

p. 24 *locked up and planed*

should read 'planed and locked up'. See note to p. 23, *supra*.

p. 27 *the exact height at which the roller should be held*

The difficulty would have been avoided if advice had been given in the use of type-high inking bearers.

p. 28 *The tympan sheet*

Professionals positioned their tympan sheets by laying a pasted sheet on the forme and then lowering the tympan onto it. In adapting this method for the amateur the present account garbles it and makes it needlessly complicated. A simpler and more certain method would be to make a lightly inked impression on a large sheet of paper previously pasted to the tympan and to position the lay pins on this paper with the aid of a ruler.

p. 30 *kerseymere cloth*

'A twilled fine woollen cloth of a peculiar texture, one-third of the warp being always above, and two-thirds below each shoot of the weft.' (O.E.D.)

p. 31 *a roll of paper is pasted on the frisket*

The usual purpose of such rolls of paper is to prevent slurring. They achieve this by forcing the frisket paper hard against the paper being printed, so preventing it from sagging and making premature contact with the type.

p. 35 *These apparatus are larger than the parlour presses*

See note to p. 5 on the sizes and prices of the various presses made.

p. 36 *the types in the four lower drawers comprise upwards of 9000 pieces*

If the Pica Roman is taken as 5000 pieces it would weigh about 18 or 20 lb; the Bourgeois Roman might have 2000 pieces and weigh nearly 5 lb; and the Bourgeois Antique and Great Primer Roman might have 1000 pieces each and weigh $2\frac{1}{2}$ lb and 8 lb respectively.

p. 36 *Bourgeois Roman*

According to the reference to this fount on the next page, it also included lower case letters.

p. 48 *a gentle pressing*

Frederick Noble, in his *The principles and practice of colour printing stated and explained*, 1881, suggests that the 'metalled sheet may be placed on a type-high board with a blanket on the top of it and pulled over in the press; this forces the metal into every part of the impression and causes perfect adhesion.' He also remarks on the need to make the work 'as flat and smooth as possible; . . . . To this end the forme, no matter what it may be, should be made ready with a hard impression, and the bringing up should be such that the impression should not "dip" in any part.'

p. 55 *a selection of about 1500 types*

Probably the Small Pica Roman and Bourgeois Roman would

have comprised 5A 5A 12a founts (accounting for about 500 types each); the Small Pica Black letter 5A 12a (say 350 types); and the Bourgeois Antique about 5A (say 100 types).

p. 55 *The hand chase*

Supplied in a painted case, with cushion, roller, ink, and inking tray, for 15*s* (see the Price List, p. 79).

p. 57 *Monotype Printing Press*

Supplied in a mahogany box, with type, black paper, and cards, for £5 15*s* (see the Price List, p. 79). It was listed, unpriced, in Holtzapffel's general catalogue for the first time in 1834.

p. 59 *camp paper*

O.E.D. defines this as 'a kind of carbon paper', and quotes John Imison, *The School of Arts*, ii, 31, London (*c.* 1790): 'To make Camp Paper, with which a Person may write or draw without Pen, Ink, or Pencil.'

p. 60 *A very useful application of Monotype Printing*

Robert Willis (1800–75) was Jacksonian Professor of Applied Mechanics in the University of Cambridge. The range of his ingenuity is nicely illustrated by two of his inventions (both stocked by Holtzapffel and Co): the cymagraph, for tracing architectural mouldings, and the odontograph, a device for setting out the teeth of gear wheels. His mode of printing architectural plans is known nowadays as stickprinting, and is taught in schools as an advanced form of potato printing.

p. 62 *Specimens of Types, &c.*

With four exceptions the types are from the Caslon foundry, and are shown in their *Specimen of Printing Types*, London, 1844. Two types (21 and 61) are from the Fann Street Foundry of William Thorowgood and are shown in his *Specimen of Printing Types*, London, 1842. Types 70 and 72 have not been identified. The ornamental material is all from the Caslon Foundry.

p. 63 *No. 1. Five Line Pica*

The actual sorts shown in the two specimens, Nos. 1 and 2, of Five Line Pica would have cost 9*s* 2*d*; they would have weighed just under $4\frac{1}{2}$ lb and would have cost 4*s* 6*d* if bought by weight from Caslon (Henry Caslon, *Specimen of Printing Types*, 1844).

Similar calculations for three of the other faces shown yield the following comparative costs.

No. 13, Pica: cost from Holtzapffel by the dozen sorts, 7*s*; cost from Caslon by weight (about 1 lb), 1*s* 7*d* (Caslon's *Specimen* of 1844).

No. 17, Bourgeois: cost by the dozen sorts, 7*s* 6*d*; cost from Caslon by weight (about $\frac{2}{3}$ lb) 1*s* 7*d* (Caslon's *Specimen* of 1844).

No. 21, Great Primer Script: cost by the dozen sorts, 22*s*; cost from Thorowgood by weight (about $1\frac{1}{4}$ lb), 6*s* 3*d* (Thorowgood's *General Specimen of Printing Types* of 1848).

The Caslon foundry itself held aloof from the trade in small founts for amateurs. 'It does not pay,' they said in *Caslon's Circular* for July 1879. It is likely however that Holtzapffel made it pay handsomely, for labour was cheap, and his prices seem to have been anything up to five times (and this in the most popular sizes) those charged by the founders.